Leonard L. Conkey

Veterinary Medicine

animal castration, surgery and obstetrics simplified

Leonard L. Conkey

Veterinary Medicine
animal castration, surgery and obstetrics simplified

ISBN/EAN: 9783337239527

Printed in Europe, USA, Canada, Australia, Japan

Cover: Foto ©berggeist007 / pixelio.de

More available books at **www.hansebooks.com**

Leonard L. Conkey

VETERINARY MEDICINE,

ANIMAL CASTRATION,

SURGERY AND OBSTETRICS

SIMPLIFIED,

—BY—

LEONARD L. CONKEY, V. S.

—SPECIALST IN—

CRYPTORCHIDE (RIDGLING), CASTRATION, SURGERY AND OBSTETRICS; INVENTOR OF
"THE CONKEY PATENT HOBBLE BUCKLE," "THE CONKEY OBSTETRICAL SET,"
"THE CONKEY INCISIOR CUTTING FORCEPS" AND "THE CONKEY DOSE GUN."

GRAND RAPIDS, MICH.
VALLEY CITY ENG. AND PRINT'G CO.
1890.

PREFACE.

In preparing this work it has been with a desire to present to the public a practical Hand Book in plain terms, giving the causes, symptoms and treatment of the diseases and accidents common to our domesticated animals. This little book is expressly adapted to the use of stock growers and horse owners, as well as the farmers in general.

For convenience I have divided this work into nine chapters, as follows :

CHAPTER I.
DISEASES—CAUSES, SYMPTOMS AND TREATMENT.

CHAPTER II.
LAMENESS—CAUSES, SYMPTOMS AND TREATMENT, WITH A FEW PRACTICAL HINTS ON SHOEING.

CHAPTER III.
VETERINARY SURGERY.

CHAPTER IV.
ANIMAL CASTRATION, AND TREATMENT OF THE DISEASES FOLLOWING IT.

CHAPTER V.
DENTAL SURGERY—THE AGE AS INDICATED BY THE TEETH.

CHAPTER VI.
THE EYE AND EAR.

CHAPTER VII.
OBSTETRICS—"DELIVERING THE MOTHER OF HER YOUNG," WITH DISEASES FOLLOWING.

CHAPTER VIII.
DISEASES OF THE COLT AND CALF.

CHAPTER IX.
VETERINARY MEDICINE, INCLUDING THE DOSE TABLE.

Each chapter is illustrated with engravings produced especially for this work. Many of them are from pen sketches by the author

The professional veterinarian has already at his command many large volumes treating on the domesticated animal. These, however, devote much space to the consideration of subject familiar to the professional man only, with the free use of technical terms, making them "Greek" to the non-professional man. For this reason, I have not referred to physiology and morbid anatomy in this little work.

The embryo of many principles advocated in this work were taken from the following standard authors: Professor Fleming, of London, Eng.; Professor Smith, of the Ontario Veterinary College, Toronto, Canada; Professor Liautard, of the New York Veterinary College, U. S. A., and Professor W. Williams, of Edinburg, Scotland. These principles have been nursed by the AUTHOR through his many years of practice with flattering results.

I sincerely hope that this work will be found useful, and that it will assist in promoting the science of veterinary medicine, surgery and obstetrics, which has for its object the benefit of man financially and the relief of our dumb animals directly.

LEONARD L. CONKEY.

BENTON HARBOR, MICH.

VETERINARY MEDICINE;

CHAPTER I.

Diseases, Causes, Symptoms and Treatment.

THE PULSE.

The pulse is the beating of the arteries caused by the action of the heart by which we are able to distinguish different diseases, and can be best felt by placing the fingers on the sub-maxilliary artery about half way between the throat latch and lower lip on the inside of the lower jaw. In the cow it is best felt on the inside of the front leg just above the knee, while in the dog it is best found on the inside of the thigh, and the number of beats in the different animals are as follows: Horse, about 40 beats per minute; cow, 40 to 45; dog, 80 to 100; while in the sheep from 70 to 80, and is found in same place as in the cow.

TEMPERATURE.

THE FEVER THERMOMETER.

The warmth or degree of heat of the animal body is an index to health and disease. In health the normal temperature per mouth is 98½, and per rectum 99½ to 100. On account of the animals biting the thermometer, the rectum is selected as the proper place to take the temperature. This is done by carefully introducing the thermometer into the rectum about two-thirds of its length, and allowing it to remain two or three minutes. The thermometers now in use are self-registering, so that there is no need of being in a hurry as the mercury will not go down until it is shaken, which must always be done before using.

THE RESPIRATION (BREATHING.)

The normal (natural) respiration (breathing) of the the horse is from twelve to fourteen times a minute if taken while standing quiet. This is done by placing the hand to the side, just in front of the hind leg, or it may be taken by the eye alone. The normal respirations of the cow are somewhat slower than that of the horse, being ten or twelve times per minute.

THE KEY TO PRACTICE.

When once you become acquainted with the natural *Pulse, Temperature and Respirations* of the different lower animals, you will have the *Key to Practice.* Any noticeable change above or below this is indicative of disease.

For example; when we have a quick, bounding pulse, say 60 to 80 beats per minute, we will give a sedative, aconite, digitalis, belladonna, etc. Then if we have a slow, weak pulse we will give a stimulant such as alcohol, amonia, S. S. niter, etc. Again, should we have an increase of temperature showing fever, we will give refrigereuts as nitrate of potash, cold water, etc. Then should we have a fall of temperature we will give stimulants such as whisky, spirits of turpentine, ammonia, etc. Thus you will see how very easy it is to treat the diseases of our domesticated animals when you have the key, *i. e.*, Pulse, Temperature and Respiration.

CONGESTION OF THE LUNGS

Is a forerunner or inflammation, or pneumonia, and is the most common disease of the lungs.

Causes—Rapid exertion when the animal is not in a fit state is the most common cause. It is a sequel of catarrh, and if a horse is put to work too soon, it may follow influenza. I have seen horses take congestion from standing in a draft of air when heated and cool off too suddenly.

SYMPTOMS.

If from fast work, the symptoms are well marked ; the animal will be sluggish, tremble in the flank, will have labored breathing, the nostrils dilated, oppressed pulse: sometimes very weak and indistinct, the lining of the nose and the white of the eye are reddened. By placing the ear to the side there will be heard a peculiar gorgling noise, the ears and legs are cold ; the breath also cold, but if the attack be a mild one the symptoms may be indistinct.

TREATMENT.

This is rather a desirable disease to treat, and generally terminates favorably. Keep the horse in good fresh air ; it is better to keep him out of doors than in a close stable; blanket well and give stimulants. Give one ounce of sulphuric ether with two ounces of whisky and a little warm water every two hours ; alternate with powdered nitrate of

potash in four dram doses until relieved; give injections, per
rectum, of soap and water, bathe the legs lightly with alcohol
and hand rub them well, then bandage them with warm
bandages. If the symptoms are relieved and the pulse runs
high give tincture of aconite in 10 to 20 drop doses every
two hours. Allow plenty of cold water—a little at a time
and often. Give a few doses of quinine as a tonic.

INFLAMMATION OF THE LUNGS.

It may occur in either the acute or chronic form. It is
not at all an uncommon disease. There are several stages of
this disease, and it may terminate fatally at any stage.

Causes.—Sudden change in temperature, improper venti-
lation; animals are more liable to it during the spring and
fall than during the steady summer and winter weather;
turning from a warm stable out to pasture, clipping and then
exposing the animal to cold drafts, etc.

SYMPTOMS.

It is generally brought on by rigor (chill), when the
shivering ceases heat takes place, that is, fever sets in; ears
and legs cold, then hot, or may assume the natural temperature; breathing somewhat affected, mouth hot and clammy, or
sticky; pulse generally oppressed and quick although they
may be full, the horse persists in standing for the reason that
the thorax cavity is larger than when lying down, giving
more room for the lungs to expand; the eyes are generally of a glassy appearance, a kind of flapping of the nostrils, a
rather heavy sighing breathing, by placing the ear to the
chest a crackling sound may be heard, bowels usually constipated, feces are usually covered with slime. If the horse

looks around noticing things a good deal it is a good sign. The horse should be allowed plenty of pure air, and if turned loose you will see him go to an open door. The breathing varies to a certain extent, but not so much as might be supposed—he breathes about 10 or 12 times a minute. By-and-by the chest or thorax begins to fill with liquid called an exudate, the pulse quickens, numbering a hundred beats per minute or even more, the flapping of the nostrils and breathing increase in rapidity, there is a brownish discharge from the nostrils (this is a bad sign), breath smells bad, persists in standing. notices nothing, and as death approaches the pulse become imperceptible and the mouth cold. If an animal whose lungs are afflicted dies in from 12 to 24 hours after the first symptoms are noticed it is most likely to be from congestion. If the horse is to recover he will lift his head, look around and commence eating. Lung fever is generally satisfactorily treated; that is, with proper treatment they generally recover.

TREATMENT.

Give from 10 to 15 drops of tincture of aconite alternately every hour with powered nitrate of potash, teaspoonful doses. You may relieve the distressing symptoms by giving tincture of opium; give from one-half to one ounce at a dose. Encourage the horse to eat scalded bran mash, but if he refuses it then give him anything he will eat, but do not try to force him to eat. When you have obtained relief then give mild stimulants such as sweet spirits of niter in milk, or whiskey in milk, etc., giving a cooling laxative diet. If a cough should threaten use opium and digitalis.

PLEURISY.

This frequently exists in connection with lung fever, then

we have pleuro-pneumonia, which is a serious disease, generally terminating unfavorably.

Causes.—The causes are very much the same as in lung fever.

SYMPTOMS.

It usually begins with rigor (chill), pulse quick and fuller than in lung fever (pneumonia). The animal seems to suffer great pain, breathes quick, and unlike lung fever, he is apt to lie down : if he coughs at all it will be a suppressed cough, ears and legs cold, or he may have one hot and the others cold and vice versa ; there will be a line along the lower end of the short or floating ribs—a tucked up appearance. If you turn the animal around a little quick he will grunt or groan : tapping him on the side with the hand will cause him to grunt and evince much pain. After a short period you may think the animal is getting better, as he may look around, breath easier, and to all appearance is better, but if you notice the pulse you will see that they are running up. This is at the time when what is generally termed dropsy sets in : the chest commences to fill with exudate (a watery substance), the belly and legs may and usually are swollen. about this stage of the disease the animal may take a little food, but you will readily see that it is not with a relish.

TREATMENT.

Clothe the body well according to the season of the year, place the animal in a well ventilated box stall, hand rub and bandage the legs, give cold water at short intervals—plenty of it if he will take it—and give the following as a drench :

Sweet spirits of niter, 4 ounces ; fluid extract of belladonna, 2 ounces ; tincture of gentian, 4 ounces ; mix, shake drams;

well together. This makes four doses to be given in one-half
pint of warm water every two hours. Try and get the animal to eat the best of food and let him have it in small quantities, and it is best to put about two ounces of nitrate of
potash in each pail of water that is offered him.

PLEURO-PNEUMONIA.

Although contagious in cattle it is not contagious in the
horse. Causes and Symptoms are about the same as in
pleurisy except that the animal persists in standing, and the
treatment differs only in less sedatives and more stimulants.

THUMPS.

SPASMS OF THE DIAPHRAGM.

Thumps is a very serious disease while it lasts, as the air
cannot be taken in in a sufficient amount to supply the lungs.

Causes.—Overwork, especially when the animal is not in
fit condition; long continued fast work, although he may be
in good condition, may produce it; a horse put to work on a
full stomach is liable: also horses recovering from any debilitating disease.

SYMPTOMS.

The symptoms of thumps are so plain that you cannot be
mistaken. The horse has a loud thumping sound as though
some one was inside pounding with a hammer.

TREATMENT.

Take of fluid extract of digitalis ½ ounce, fluid extract of
belladonna 1 ounce, alcohol 4 ounces. Mix and give two

tablespoons full every two hours in a half pint of water until relieved.

INFLUENZA

Is a very common disease in the United States. It is a highly febrile disease and affects different organs of the body, the throat and lungs being more often affected than any other organ. When the liver is affected the white of the eyes have a yellow cast, and occasionally we have actual blindness from influenza. The whole mucous membrane (inside lining) of the intestinal canal is more or less affected.

Causes.—Our best authorities, after discussing the probable cause of influenza, say that whatever the cause is it has not as yet been detected, and my idea is that the specific germ must be due to some atmospherical condition and travels through the air. As to the contagiousness of the disease, there is a great diversity of opinion. It might be best to keep the sick away from the well ones, but the same predisposing cause that gives it to one horse will or may give it to a number at the same time. We find more horses affected in bank or basement barns with damp, wet floors than in any other stable. Stabling cattle in the same room with horses appears to favor the development of influenza poison. Influenza is most prevalent in the autumn, winter and spring months, but I have seen it in its worst form in the summer months. The young animals suffer most, although the aged are not exempt and often die from influenza.

SYMPTOMS

Depend much upon the organ affected. About the first symptons are dulless, lagging in the harness, sweat easily

and pant or breathe quick on the slightest exertion, coat staring, mouth hot and dry, cough easily excited by grasping the
throat, bowels usually costive and feces covered with mucous
although the bowels may be looser than common, but this is
not the rule. The pulse is quick, and in looking over my
register find that the average in one hundred cases are pulse
60, respiration 20, temperature 104. In some cases we have
an intermittent pulse showing that the disease is working on
the nervous system. In such cases the breathing is not so
much affected. Spinal complications are produced in this
way. In other cases the breathing is much affected, the
throat very sore, and by placing your ear against the windpipe
near the breast you will hear a peculiar, unnatural sound.
The legs and ears change in temperature very much, now hot
and in an hour cold and vice versa. The general temperature
will run up to 105 or 106. The eyes are sometimes of a reddish or pink color owing to congestion. A light colored discharge from the nostrils is always a good symptom, but if it
is of a rusty brown color it is a very bad sign. If the pulse
is very changeable it is also a bad sign. Influenza may terminate in inflammation of the bowels (enteritis) and deaths.
Any of the secreting glands may be affected causing dropsical
swelling of the legs, belly, sheath, udder (bag of the mare),
eyelids, nose, and the whole face and head are sometimes
wonderfully swollen. In the first stages this is not a bad
symptom, but if it appears in the last stages it is to be re-
garded as a serious case.

The patient usually persists in standing although he may
lie down, which increases the labored breathing, but if he
seems to by lying easy allow him to remain as it affords much
relief. This disease has a tendency to affect the joints (arthritis); or it may terminate in (laminitis) founder, rheumatism,

etc. Thus it will be seen that we have various forms of influenza.

TREATMENT.

Give plenty of pure air. This we wish to impress upon your mind as an important part of the treatment. Clothe the body in accordance with the season of the year; hand rub the legs and apply dry bandages. In summer a light sheet should be thrown over the horse to keep the flies off. Allow plenty of cold water, and it is a good policy to keep a pailfull where he can reach it at all times, as he will often take but a swallow and apparently wash out his mouth which seems to refresh him. Influenza will run its course in spite of medicine; all we can do is to assist nature to throw off the poison, and for this you will give, in mild attacks, powdered chlorate of potash in two dram doses three or four times a day, liquor acetate of amonia, two ounces two or three times a day. Give a scalded bran mash once a day; give plenty of strong, nutricious food, if he will not eat one kind offer another, but do not force food upon him as you are liable to destroy his appetite; offer a little at a time and often. In severe cases you will give powdered nitrate of potash instead of chlorate, in from two to four dram doses every six hours; tincture of aconite, tincture of digitalis, of each one ounce, give of this 10 or 15 drops every six hours, liquor acetate of amonia two ounces every six hours, which will bring one of the medicines every two hours alternately. Give rectinal injection of raw linseed oil, flour gruel, oatmeal, etc. This is absorbed into the system furnishing nourishment as well as acting as a laxative to the bowels. Apply the following liniment to the throat: Mix aqua amonia, spirits turpentine and raw oil—equal parts; make two or three applications at intervals of 24 hours. When the animal begins to improve, which will be indicated by the eyes clearing up, the appetite returning, you will stop the aconite and give in its place one tablespoonfull of the following: Iodide of potash four ounces, water one pint. About this time you will discontinue the liquor acetate of amonia. The swelling of the legs and throat will gradually subside, Do not put the animal to work too soon, and when you do commence work be very careful for a while.

INFLUENZA OF THE DOG.

During the seasons when influenza rages as an epizootic we find many dogs affected.

SYMPTOMS.

The dog is rather sluggish, lying around, appetite poor or gone, throat swollen, tears running from the eyes, and perhaps a little matter in the corners of the eyes.

TREATMENT.

Nitrate of potash, one-half dram : quinine, two grains; to be given three times a day before meals. If the dog should be very weak give an occasional dose of whisky.

INFLUENZA OF THE SHEEP.

Sheep are occasionally affected and we have seen quite large flocks go blind from the effects of influenza and remain so for a few days and then their sight would gradually return. There was a watery discharge from the eyes, a slight cough, appetite impaired.

TREATMENT.

Give powdered chlorate of potash one dram, quinine five grains, three times a day. Give laxative food, carrots, turnips, potatoes, etc.

INFLUENZA OF THE HOG.

Hogs are also subject to this disease.

SYMPTOMS.

Do not differ much from those in other animals.

TREATMENT.

Powdered nitrate of potash, one-half to one dram; powdered nuxvomica, five to ten grains given in the slop three or four times a day to each hog of 75 to 100 pounds; larger ones more and smaller ones less than this dose. In conclusion I will say that influenza is rather a satisfactory disease to treat, and with proper treatment the animal is as good as ever. There are, however, a few cases of influenza that the swelling of the throat is so bad that the animal may die of strangulation. The only remedy for such a case is called *Trachea Otomy,* (See Index), which means to cut a hole into the windpipe, thus forming an artificial opening through which the air reaches the lungs. In the hands of a skilled operator this is a very simple operation.

STRANGLES

Is a name given to what is generally called the old fashioned horse distemper.

SYMPTOMS.

About the first indications are manifested while drinking

by the water running out of the nose, appetite impaired, slight soreness in the throat and in the angle of the jaw, the animal will carry his head in a peculiar, unnatural way, saliva runs from the mouth; coat staring, tumors will soon form in the angle of the jaw, or the tumor may possibly be the first sign of the disease. There is usually a discharge from the nose. The breathing is considerable affected, which may not be in proportion to the size of the tumor, as the swelling or enlargement may be internal, and when affected with strangles and influenza at the same time there is great danger of suffocation, when you will have to resort to *Trachea Otomy*, for which (See Index.) Strangles usually runs its course in from 10 to 15 days, and in 20 to 30 days the animal is ready for work again. We sometimes have all those symptoms without the swelling of the throat, and in a few days we find a swelling in the groin, on the shoulder, the hip or hock and various other parts of the body. This is called

IRREGULAR STRANGLES.

These tumors are the results of fever. Occasionally the animal will exhibit signs of great pain and lameness in one hind leg for several days before the forming of a tumor, and you are at a loss to know just what the ailment is. After an abscess forms the lameness and pain generally subside; the patient often becomes a mere skeleton

TREATMENT.

Take of powdered nitrate of potash four ounces, powdered chlorate of potash four ounces; mix. Give one table-spoonful three times a day on the tongue or in the feed, at the same time apply aqua amonia, terpentine and oil, equal parts,

to the throat or swelling, rubbing in well two or three times
a day, until relieved or an abscess forms, which must be opened,
making a large opening and poulticed.   If the animal appears
weak give one table-spoonfull of powdered gentian three times
a day on the tongue.   Should it be unable to eat you must give
rectal injections of oatmeal gruel, or raw lineeed oil.   This is
absorbed and assists in sustaining life.

CATARRH

Is a disease of the mucous membrane (lining) of the nose
A congestion takes place, the irritation gives way to a watery
discharge, which changes to white and then yellow in accordance with the severity of the disease.

Causes.—Standing in a draft, cooling off too soon, change
in temperature, etc.

SYMPTOMS.

At first a little dull or lazy, may not eat well, pulse is
not much altered, slight raise in temperature.   After the congestion passes off the discharge from the nose follows, and
may be profuse, and as we have a discharge from the nose in
other diseases such as influenza, strangles, glanders, etc.   You
must watch the pulse and temperature closely.

TREATMENT.

Attend to the general comfort of the animal.   A clean
box stall well ventilated.   Give nitrate of potash from one to
one and one-half tablespoonsful three times a day with plenty
of good, strong food of a laxative nature: do not allow the
animal to run down in condition.   If a cough is present bathe
the throat and angle of the jaw with aqua amonia, turpentine

and oil, equal parts. If the discharge from the nose continues give chlorate of potash in tablespoonful doses three times a day for a few days, then give powdered iron sulphate, one teaspoonful twice a day.

LARYNGITIS.

Is an inflammation of the throat and sometimes terminates fatally in a short time.

SYMPTOMS.

Swelling of the throat hard and painful to the touch. If the pulse is very fast and the animal cannot swallow you have a very bad case; the breathing hurried, bowels costive, urine scanty. In a day or two there will be a discharge from the mouth, which in the first stage is a good sign. The animal generally recovers in from a week to ten days, but the horse should not be put to work for some time as it is liable to terminate in roaring.

TREATMENT.

Plenty of pure air and cold water are very essential; clothe the body according to the season of the year and bandage the legs. Then give internally: Powdered cholrate of potash, four ounces; powdered nitrate of potash, two ounces; powdered ipecacuan, two ounces; mix. Give one tablespoonfull three or four times a day on the tongue; also give one teaspoonful of fluid extract of belladonna three times a day, one or two hours after giving the other medicine. *Externally* you will apply the following to the throat and to the angle of the jaw, rubbing it well up toward the ears two or three times a day : Aqua amonia, spirits turpentine and olive oil, each two ounces. Mix and apply as directed.

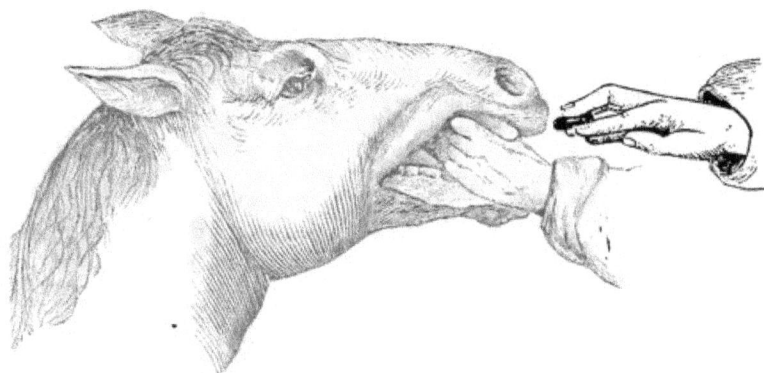

GIVING A BALL.

First untie the animal and have an assistant stand on the right side with his left hand on the nose and the right hand on the back of the neck or crest. The assistant should endeavor to keep the head straight with the body. You will take the ball or pill in the fingers of the right hand as shown in the engraving. Then with the back of your left hand upwards gently work your fingers between the lips, touching the roof of the mouth. This will cause him to open the mouth, when you will grasp the tongue drawing it out a little, and give the hand a quarter turn. This brings the tongue under and over your hand, as shown in the drawing, which prevents the animal from closing the mouth. You will now pass the ball into the mouth, placing it back over the root of the tongue. Withdraw the hand and let go the tongue, holding up the head for a moment. A little practice will enable you to give a pill or ball before you could possibly get a drench ready.

PHYSIC BALL.

Barbadoes Aloes	8 drams
Calomel	1 "
Powdered Ginger	2 "
Powdered Capsicum	1 "

OR

Barbadoes Aloes	8 drams
Powdered Nuxvomica	1 "
" Gentian	2 "
" Podophyllin	1 "

To either of the above you are to add molasses, glycerine or common soap enough to make a stiff mass, and give all at once in a ball.

GIVING A DRENCH.

First get the drench, an old chair or box, a common pitchfork and a piece of cord eighteen inches long; tie the ends together forming a loop. Put this in the horse's mouth around the upper jaw, then with the pitchfork fixed in the loop raise the head just high enough so that the medicine will run down the throat. If you hold the head too high the animal cannot swallow so well. Get up into the chair with your left hand holding the halter, which prevents shaking the head. Take the drenching bottle in your right hand, place the neck of the bottle just inside of the lips and allow a little of the medicine to run down into the mouth; do not push the bottle down into the mouth as far as possible, this not only irritates the animal, but endangers its life by the breaking of the bottle and swallowing the glass. Back the horse into a

stall or a corner that his hips may rest against something
while giving a drench. This prevents him pulling back and
breaking away. Do not jerk and abuse him, but go slow and
careful, and you will get through more quickly.

DRENCHING THROUGH THE NOSE.

Never do this yourself, nor allow any one to do it for
you. We cannot too strongly condemn this barbarous practice of pouring medicine down the poor creature's nose.

THE PHYSIC DRENCH.

To either of the above add boiling water to dissolve, then
add a little cold water to cool and give all at once as a drench.
These prescriptions are intended for horses of mature years,
and large horses on dry feed may require more, while small
horses may require less. The weanling colt will require
about one-eighth of the above dose, the yearling one-fourth,
the two-year-old one-half, and the three-year-old three-quarters of the above doses. You must mix a little common
sense with those remedies and you will get excellent results.

GENERAL REMARKS ON PHYSIC.

A scalded bran mash given twelve hours before a physic
will increase and hasten its action. You should always give
a scalded bran mash immediately after a physic and withhold
the hay for twelve to twenty-four hours. Give pure, fresh
water freely, a little at a time and often, to all animals after
administering a physic. Should a physic fail to operate in
from 24 to 36 hours it is best to give one-half pint of good
whisky and one-half dram of fluid extract of nuxvomica as a

drench, and it might be well to give a little walking exercise about this time.

HOW OFTEN TO REPEAT A PHYSIC.

We will suppose that you have given the physic ball or drench, waited 36 hours and then gave the whiskey drench. You will now wait 36 to 40 hours longer when, if the bowels are not open, you are justified in repeating the physic ball or drench.

SUPERPURGATION.

THE BAD RESULTS OF PHYSIC.

Superpurgation does not always depend upon the strength of the dose, but largely upon the condition of the animal at the time of its administration. In some animals a very small dose has produced superpurgation. Horses suffering from colds, influenza, etc., are easily purged. A full dose of aloes (eight drams) operating quickly is seldom followed by any bad results, while the same amount divided into two doses given some hours apart, is apt to be followed by serious results. In the first instance the quantity by its strength insures its own expulsion, while in the second instance the aloes is absorbed into the circulation exciting the bowels to an extent sufficient to produce superpurgation.

So long as the animal remains moderately lively, the pulse not much affected, the countenance natural, the appetite not much impaired, it is not necessary to take any active measures to restrain the purging, which is only the natural response of the intestines to the action of a physic, therefore it is dangerous to check it too suddenly. Should you be alarmed give raw flour gruel to drink, and do not allow the animal to drink too freely of anything until the bowels are set. Colicky pains are to be treated with one-half ounce doses of tincture of opium, great care being taken that the purging is not checked too suddenly. Should the purging continue after 24 hours it is to be treated the same as *Diarrhœa*, which see.

HEAVES, OR ASTHMA.

We will not enter into a lengthy discussion as to the causes and symptoms of this disease; suffice it to say that it is incurable, but the symptoms may be relieved by the following .

TREATMENT.

Give a good physic composed of the following: Barbadoes aloes, one ounce; powdered nuxvomica, one dram; powdered gentian, two drams. Make in a ball with glycerine or give in a drench. After the physic works off feed the animal good, clean food, bright hay, oats and corn Wet the hay and feed it in small quantities, and grain enough to keep the animal in condition. To relieve the distressing symptoms try one of the *heave remedies*. (See Index). Should you get no relief from the first remedy, after a week or two try another, for heaves are not always due to the same causes, and the different forms require different treatment.

ECZEMA

Is an eruption of small vesicles on the skin, and is essentially a "hot weather" disease. It is often mistaken for mange. The first symptoms are dryness and itching about the neck, ears, back and tail. Then little vesicles or blisters appear. These sometimes burst and discharge a yellowish fluid, but more commonly they are absorbed, leaving dry scabs. The animal rubs himself against any convenient object until at times there are large raw sores along the neck or back.

TREATMENT.

Many authors say that eczema is very difficult to treat, and you are not likely to effect an entire cure. I think, however, that if the proper treatment is applied, at least nine-tenths of those affected can be cured. First clip the animal, then scrub him with soap and water, rub him dry, and sponge the affected parts with the following: Take one pint of boiling water, add corrosive sublimate, two drams, allow this to cool and use just enough to moisten the sores once a day for three days. Then use white lotion as a wash for a few days, then use carbolic acid, one ounce, water, one pint, for a few days, then return to the white lotion again.

INTERNAL TREATMENT.

First give a physic ball or drench followed by a scalded bran mash, (withholding the hay for 24 hours). Then take iodide of potash, four ounces; water, one pint; mix. Give one tablespoonful three times a day in the feed until all is given, after which you are to prepare powdered starch, one pound; arsenic, three drams. Mix thoroughly and give one tablespoonful in the feed three times a day until all is given. Repeat the physic once every two weeks until you have given three or four doses.

LYMPHANGITIS,

Monday morning fever, shot of grease, water farcy, milk leg, etc. However, I shall call it lymphangitis when speaking of it in this chapter, giving you the most common causes, symptoms and treatment. This disease is common among heavy work horses, often brought on by allowing them to remain in the stable over Sunday, giving the usual amount of feed without the usual exercise. A larger amount of nutritive material is formed than can be taken up, which sets up an irritation, and on Monday morning you find the affected limb swollen and perhaps hanging pendulous, hence the name Monday morning fever. Direct or indirect injury to the groin, slipping in the act of getting up in the stall; a prick or gravel in the foot is also liable to produce it. A newly purchased horse is liable to it from the unusual amount of nourishing food given with a view to make him a little nicer. The first symptoms are a rigor (chill) which may not be observed. A rise of temperature to 103 or 104°, breathing quick, pulse about sixty to eighty beats per minute, full and bounding. Appetite at first impaired, but usually returns in a day or so, bowels costive, urine highly colored, the desire for water usually intense, the limb is swollen, sometimes all around and sometimes on the inside only, a sort of a corded line from the groin to the hock joint. If you pass your hand along this cord it causes the animal to lift his leg quickly. The animal usually stands, but he may lie down. Occasionally you may find a case where the animal looks at its sides as in colic. The leg will regain its natural size if properly treated. From repeated attacks the leg is enlarged, the lymph becomes organized with, and is a part of the leg, and cannot be got rid of.

TREATMENT.

First give a strong purgative for the purpose of setting the excretory organs to work, that they may carry out the worn out matter which has been retained in the system. Barbadoes aloes, eight drams; powdered nuxvomica, one dram; powdered ginger, two drams; powdered calomel, one dram; make into a ball or dissolve in one pint of boiling water, and give all at once. A larger horse will require about one-third larger dose. Then get one-half pound nitrate of potash and give one tablespoonful every four hours on the tongue; also get of fluid extract of aconite, one ounce; water, one pint; mix and give one tablespoonful every four hours alternately with the potash. This will bring one medicine every two hours. Give warm water injections per rectum every time you give a dose of medicine until the bowels move freely. If the bowels do not move freely in from 24 to 36 hours give one pint of raw oil containing two ounces of spirits of turpentine as a drench, and repeat the oil once a day until the bowels move freely. The diet must be carefully regulated. Withhold hay or other coarse food until the acute symptoms begin to subside Give a scalded bran mash morning and evening if the animal will take it, if not, give a quart of oats three times a day for a day or two, gradually increasing to the usual feed.

EXTERNAL TREATMENT.

If there is much pain bathe the leg three or four times a day with the following: Tincture of opium, four ounces; whisky, eight ounces. Bathe freely and hand rub for at least twenty minutes. As soon as the acute symptoms begin to subside you are to give a little walking exercise, but under no circumstances are you to move an animal suffering from an acute attack of lymphangitis. Rarely we meet with a case that in spite of medical aid an abscess will form on the leg, which will break and discharge puss. This is to be opened up with the knife that the puss may escape freely, then inject a little *White Lotion* once a day, bathing the leg freely with the lotion at the same time until it is healed.

SURFEIT.

This disease comes very quickly, and is generally the result of faulty feeding; it may disappear as quickly as it comes.

SYMPTOMS.

Pimples or blotches of various sizes raise up on different parts of the body. This may follow laminitis (founder). It may be caused by drinking cold water when the animal is hot, or cooling off too suddenly. Overripe food is liable to produce it.

TREATMENT.

Give a *Purgative Ball* and in most cases as soon as the purgative acts the pimples will disappear. Give sweet spirits of niter, one or two ounces, once a day, or you may give powdered nitrate of potash, one tablespoonful; powdered camphor gum, one teaspoonful, two or three times a day. Change your feed to a laxative diet.

ITCH.

Itching about the mane and tail. You sometimes have a horse that will rub his mane and tail until it is nearly all worn off. Upon close examination you will find no parasite as in mange. This may be the result of improper care, but is more likely to be the result of poor food, and may be a symptom of worms, when the tail alone is rubbed. This, however, is an exception, and not the rule. Some say that it is a symptom of lampas (an inflamed condition of the bars of the mouth), but this is not true. (See Lampas).

TREATMENT.

Give the turpentine drench, followed by scalded bran mash and laxative food. Externally wash the affected parts with strong soapsuds and rub dry, then bathe with the aqua corrosive wash once a day for three or four days; wait a week and repeat.

MANGE.

Is due to parasites. It is an eruptive skin disease common to all domestic animals, and transmissable from one species to another. The parasite or insects are of different kinds, varying in size and shape. Some burrow under the skin; others just hold to the skin. Animals in poor condition are more liable to mange, but those in good condition will be attacked. The means of communication are various; by harness, saddle, brushes, etc.

SYMPTOMS.

Are indicated by the animal rubbing himself, the hair comes out easily, it is generally about the mane and tail. Eczema sets up more irritation than mange; the symptons are similar, but eczema spreads more quickly. If you have any doubt as to the ailment, use the microscope. This is not a frequent disease among American horses, although I have seen a number of cases among the Texas ponies.

TREATMENT.

This is somewhat tedious. First wash the affected parts with warm soapsuds, rub moderately dry, then apply one of

the following remedies: Carbolic acid, (full strength), one
ounce; water, one pint: or, corrosive sublimate, two drams;
water, one pint: or creosote, one ounce; sulphur, one ounce;
lard, one pound, rubbed well together: iodide of sulphur, one
ounce: cosmoline, four ounces. In bad cases clip the hair
from the affected parts and change remedies every two or
three days. Give internally the following: Arsenous acid,
three drams; prepared starch, one pound. Mix well and
give one tablespoonful three times a day in the feed, which
should be of a laxative nature. Groom well to keep the pores
of the skin open. The same treatment is applicable to cattle,
somewhat stronger; to sheep, a little weaker—and the carbolic
acid is perhaps the best treatment for sheep. Mange affects
the back, and eczema the belly of the dog.

TREATMENT OF MANGE IN THE DOG.

Clip the hair off of the affected parts, wash well with soap
and water, rub dry and apply carbolic lotion or blue ointment. Do not cover too large a surface at one time, as it is
apt to be absorbed and produce poisoning. Cover a small
surface, then in a few days a little more until you have covered the afflicted surface, then after a few days wash with
warm water.

LICE—POULTRY.

Lice gets up considerable irritation; more than mange or
eczema; the horse is restless, rubs himself almost incessantly.
They may be found at all seasons of the year.

TREATMENT.

Remove the cause by changing stables. If this cannot
be done move the chickens and whitewash the stable, clip the

horse nicely, then brush and groom him well if in winter, but if in warm weather wash with strong soapsuds, rub moderately dry, then moisten with one of the following lotions : Stavesacre seed, one ounce ; white helabore, one ounce ; water, one gallon ; boil down to one quart. or, powdered cocolus indicus, one-half pound ; vinegar one gallon ; place over a slow fire until it comes to a boil ; carbolic acid, two ounces : water, one quart. Just use enough to moisten the body, and repeat once or twice at intervals of four or five days. Another good remedy for cattle is equal parts of oil of tar, benzoin and linseed oil rubbed along the back, and about the head and ears. Be sure to shelter from cold rains and bad, stormy weather during treatment.

SUNSTROKE.

This is occasionally met with in all animals. It is a congestion of the blood vessels of the brain.

SYMPTOMS.

The animal will show a staggering gait or may fall, struggle for awhile, then be quite still for a time from complete loss of power, or he may try to rise and fall. In this way he is liable to do himself great injury. Pulse quick and very weak, breathing stentorious, generally make no resistance when you try to raise him.

TREATMENT.

If in a sleepy condition you may apply cold water by means of wet cloths, or pound up ice and put into a bag and apply to the head. Keep the body warm with stimulants, use warm water and turpentine, and if the animal can swallow give nitrous ether, one ounce ; water, two ounces.   If the animal cannot swallow then give rectinal injections of ether. This works nicely where there is great prostration.   If there are signs of returning consciousness there are hopes of recovery.   As soon as he will take it give clear, cold water.   As soon as you think he can stand help him up, then give a light purgative followed by bromide of potash in tablespoon doses; once or twice in two or three hours.   Watch closely for sometimes the case may be apparently doing well when a relapse soon carries the animal away.   Keep him from the rays of the sun and give rectinal injections hourly until the bowels are moved.

STAGGERS—VERTIGO—EPILEPSY—MEGRIMS.

This may be the result of temporary congestion of the brain, or anything that will interfere with the flow of blood, or it may be a sequel of heart disease, or it may be due to engorgement of the stomach and chronic indigestion; but the true causes are generally obscure; probably a morbid condition of the brain, very hard to account for.   On post mortem tumors have been found in the brain.   Nervous animals are prone to the disease.

SYMPTOMS.

The horse attacked suddenly staggers or rears and falls to the ground.   The animal may rise in a few minutes apparently as well as ever.   In some cases you will have premonitory symptoms such as drowsiness, a peculiar working of the ears alternately forward and backward ; in other cases these symptoms are absent; do not confound choking with this disease. Such horses are very dangerous, as they may become unmanageable at any time, endangering the life of the driver.

TREATMENT.

If the animal be in good condition take about three or four quarts of blood from the jugular vein; make a large opening that the blood may flow freely. After pinning up the wound give a gentle purgative followed by bromide of potash, six drams; bromide of amonia, two drams; warm water one-half pint; mix. Give all at once as a drench. Repeat this dose once a day for several days, feed on soft, laxative food, and repeat the above whenever you see the slightest signs of approaching disease.

TRISMUS*—"LOCKED JAW."

TETANUS. †—"LOCKED BODY."

This is a very dangerous and fatal disease; most animals die, although a few cases recover. The most common causes are punctured wounds in the feet by picking up nails, etc., but occasionally we have tetanus without any visible cause. Again it will develop itself from very trivial causes. I had one case following a seaton (commonly called a rowell) in the shoulder for the treatment of sweeney. Some writers say that it may follow blistering, docking, etc. Castration is another common

---

† *Tetanus*, "Locked Body."—A disease which consists of a permanent contraction of the muscles without alteration or relaxation; rigidity and immobility of the limbs and body.

* *Trismus*, "Locked Jaw."—A permanent contraction of the muscles of the jaw causing a partial or complete closure of the mouth.

cause, and makes its appearance about the eighth or tenth day and occasionally later. The causes of locked jaw following castration differ. Cold rains, lying on damp ground, standing in a pool or pond of water, are the most common causes, while a bath is liable to produce it. Some claim that there is a mechanical pressure brought upon the nerve by the healing of the wound.

SYMPTOMS.

The symptoms are usually very plain, especially if you have trismus and tetanus combined. If you have *trismus* alone the animal will move about as in health, except that he will not eat, the head is somewhat extended and you are at a loss to know just what the ail is. Put your hand under the chin and raise the head carefully when you will notice the haw cover the eye. If you have *tetanus* alone, the animal will exhibit a stilty, jerky action when compelled to move. The muscles standing out bold, firm and rigid. When left alone undisturbed the animal will lower its head to eat and drink, but the slightest noise, as clapping the hands, snapping the finger, will cause the poor sufferer to go into convulsions.

TREATMENT OF TETANUS—LOCKED BODY.

First remove the cause. If it be a nail or gravel in the foot or a suppurated corn open with a large opening, then apply the *oil cake meal poultice*. Keep the poultice warm by pouring on warm water three times a day. Place the sufferer in a loose, dark box stall away from other animals and noise. When approaching the patient do so carefully to avoid excitement. Give at once a laxative ball or drench. Barbadoes aloes, eight drams; powdered nux vomica, 1 dram; powdered ginger, 2 drams; soap or molasses to make into a ball, or you

may dissolve in hot water and give as a drench, then give the
following three times a day in the feed or on the tongue:
Quinine, six drams; powdered bromide of potash eight
ounces; powdered gentian, four ounces; mix. Give a tablespoonful three times a day. Feed scalded bran, oats and
other loosening food, and every night at bedtime give chloral
hydrate, four drams; bromide of amonia, two drams; water
and raw linseed oil equal parts to make one pint; dissolve the
medicine in the water first, then add the oil. Give as a
drench. Allow all the water he will drink, chilled in cold
weather, a little at a time and often.

TREATMENT OF TRISMUS—LOCKED JAW.

The treatment consists in taking a rubber tube, about
five feet in length and half an inch in diameter, one end of
which is drawn over the neck of a long two-ounce bottle, containing six drams of ether, and the other end is introduced
into the rectum about eight or ten inches. The bottle is then
placed in a can of boiling water, and the ether is slowly evap-
orated, which will be accomplished in about 15 or 20 minutes.
This procedure is repeated four or five times a day; and to
improve the treatment, half an ounce of chloral hydrate is
given once a day in a liquid bran mash. Besides this, the
patient is to be placed in a commodious box stall and kept
completely dark. The result of the treatment will soon show
itself. Under the constant influence of partial anæsthesia, the
temperature will soon commence to fall; the pulse falls below
normal in the course of two or three days, the spasmodic
contraction becomes gradually less, and in 20 to 25 days we
may look for recovery.

COLIC IN HORSES.

The study of colic is one of the very foremost in the

Veterinary profession. The frequency of its appearance, the
severity, the great mortality and pecuinary loss, together with
the difficulty of prevention, justifies the statement and fully
explains the prominence of its claim to the consideration of
the grower as well as the practitioner.

GENERAL SYMPTOMS OF COLIC.

It comes on very suddenly if not instantaneously. At
times the attack begins by the patient manifesting a degree of
dullness. He looks at his flank and stands back to the length
of his halter. But whether appearing suddenly or developing
slowly, the patient becomes more or less restless. He paws
and stamps, twists his body, kinks his tail, bends his knees,
brings all his feet together, makes the attempt but does not
lie down, or he may lie down but soon rises again ; he may
roll over or balance himself on his back; or rest flatly on his
side. In some forms of colic the animal, when lying down,
expresses his suffering by moaning or grunting loudly. In
all cases the countenance is anxious and contracted, the nostrils dilated, the eyes widely opened and fixed with an expression indicating the pain he suffers. His movements and struggles are more or less violent according to the degree of pain
he endures. In some animals these symptoms are of long duration and persistency, and there is no intermission in the restless motions; the twisting of the body, the stamping of the
feet, the lying down and getting up, and other indications of
the pain which has attacked the vitals of the tortured victim,
and for which he is vainly seeking relief in his contortions
and struggles. But occasionally the disease is marked with
distinct intermission, more or less characteristic. During the
remissions the patient remains quiet in his stall or stretched out
upon his bedding, at times grunting uneasily, as an expression

of the suffering he is enduring. Every individual horse, however, has his own peculiar form of attack, and his special mode of exhibiting his distress, and the general manifestation of the disease will, therefore, be modified and diversified ac-according to the individuality of the patient. The breathing is usually quickened or hurried in colic, and remains so, in various degrees, until the end of the attack, fatal or favorable. But on the other hand, notwithstanding what has been written on the subject, the circulation at the beginning of the attack seems to be reduced and greatly impaired. Arterial action is depressed, the pulse hard, small and irregular, often below normal. During this period the visible mucous membrane is pale. The heart, for some unknown reasons, remains indifferent to the existing morbid conditions, and even diminishes its action, in consequence of which the circulation is weaker. The temperature of the surface of the body is lowered. This is most marked in the extremities. At a later period the heart again accelerates its actions and its contractions become strong and repeating, though the pulse generally remains weak, small and thready, and towards the final struggle for life becomes imperceptible. There is often profuse perspiration from the beginning of the disease, which indicates a favorable termination. In cases of fatal termination this shows itself toward the end and is a cold perspiration, wetting the whole body and often running off in streams.

Certain forms of colic recover either by treatment or by instantaneous natural reaction. This is shown by the expulsion of wind, stools and urine. Preceding or immediately after these expulsions, the animal shakes himself and changes instantly from the peculiar and intense expression of pain he has endured.

Occasionally, and even with the best care and treatment,

the symptoms continue to increase in severity, the animal throws himself around more and more rapidly, and the pain becomes more and more marked at each renewal. The animal ceases to notice anything and recklessly throws himself down regardless of pain or possible injury. The respiration (breathing) increases more and more; soon a deceptive appearance of improvement presents itself, the patient seems to become more calm ; he stands back the length of his tie strap, his legs are wide apart, his features are still characterized by an expression of the agony he has endured, he is pulseless, and the coolness of the body is more marked. He lies down—now in a more careful manner—stretches his legs, and with a few slight convulsive efforts death ends the scene. The battle is over and the victim has died, exhausted by the excrutiating pain which has tortured him.

Even in slightly severe cases of colic all the functions of the intestinal canal and blander are stopped. There is paralysis of the muscular coat of these organs, consequently the passage of the food, and of the gases from the stomach and intestines is arrested while *urinating* is also suspended. It is a remarkable fact that during stomach or intestinal colic the functions of the bladder are entirely suspended, the patient often stretching himself to urinate, but failing.

Violent pains, whether continuous or intermittent, cause struggling more or less, causing the animal to breathe fast; pulse at first normal or perhaps slow, is now quick and thready or even absent, according to the termination ; bowels constipated or expulsions of gases and a few droppings; strong efforts to urinate, generally abortive ; a sudden disappearance of the colic when the attack ends favorably ; a gradual increase of the symptoms followed by a deceptive improvement is a sure forerunner of death.

The diagnosis of different colics, or more properly the affections of which colic is a symptomatic expression, is by common consent of the best practitioners and standard authors exceedingly difficult, if not impossible, in some cases to determine. "If it is true that in cases of colic we are frequently unable to go back from the symptoms to the determining cause, and to fix positively the nature of the cause. It is at least also true that it is possible, by careful study; of all the characteristics which belong to colics, to form a diagnosis which, if not positive, may be at least strongly probable of the nature of the alteration which causes it."

It results from these well founded considerations that the diagnosis is in fact really possible in a certain proportion of cases, although not in all. Just as we sometimes encounter cases of lameness that baffles our ingenuity to discover the seat or define the cause.

The practitioner who encounters one of these incurable cases should never decline attempting a diagnosis. He should study the various attitudes of the patient and their different movements and actions, which become of indispensable value in aiding us to make our diagnosis full and correct.

Colic has been divided into six classes. The first class includes the nervous and spasmodic forms of colic and generally manifests itself sometimes after the animal has eaten or drank and often on returning from a journey. They are produced by an irritation of the sensitive nerves of the stomach and intestinal mucous membrane. The pains produced are partially continuous and vary in intensity with a duration of a few hours. When exercised motion is not painful, and in some instances affords relief. The diagnosis is easy and termination favorable, except in cases of complication.

TREATMENT.

Place the animal in a well ventilated, loose box stall with a deep, soft bed, and give at once the following: Raw linseed oil, one pint; spirits turpentine, two ounces; carbonate of amonia, two drams; mix. Give all at once as a drench; if no better in 30 or 40 minutes you are to take of hyposulphite of soda, four ounces; chloral hydrate, four drams; warm water, one-half pint; mix. Give all at once. Repeat this every hour or two until relieved. Then give the following for two or three days: Powdered nuxvomica, two ounces; powdered ginger, four ounces; powdered gentian, two ounces; mix. Give one tablespoontul three times a day in the feed, or on the tongue.

SECOND CLASS.

The second class includes wind colic, which is common to cribbers and heavey horses, and usually occurs after a hearty meal. A spontaneous cure often occurs, or it is relieved by treatment until the day comes when some complication is present and the animal dies. This class of colic is recognized by the bloating, having a hollow sound or percussion. The history of the patient will greatly assist in arriving at a proper diagnosis. This class of colic also includes simple indigestion, such as are complicated with overloading the stomach; the varying degree of intensity is indicated by an anxious expression of the face, nostrils dilated more or less, distension and heaviness of the abdomen, pain and hesitation in walking, falling down heavily and complaining loudly, gapping more or less frequent, bowels constipated, urinary functions suspended. In indigestion with overloading the diagnosis is quite easy. The prognosis must be a guarded one on account of the serious complications with which they are likely to be associated. These forms of colic are likely to follow hearty and hoggish eating or drinking.

TREATMENT.

Take of good whisky one-half pint; bicarbonate of soda, one ounce; carbonate of amonia, two drams; warm water, one-half pint; mix. Give all at once as a drench. As soon as you have given this you must. without delay, give a physic ball or drench. (See Index). Should the pain continue unabated, in from 15 to 30 minutes you are to give chloral hydrate, four drams; carbonate of amonia, two drams; bicarbonate of soda, four drams; warm water, one-half pint ; mix. Give all at once as a drench, and repeat this as often as may be necessary to keep the animal quiet. Should the gases continue to form, which is indicated by the distension of the abdomen, you are to tap the animal, using the *Trocar and canula.* (See Index.) This is done by puncturing the colon on the right side, mid-way between the last rib and the point of the hip at the most distended or prominent point. Dr. Rose, of Grand Rapids, Mich., says he has tapped many hundred horses during his practice in the city with good recoveries. Some of them were tapped 13 and 14 times in 24 hours and recovered without even an abscess forming, which sometimes follows the wound made by the trocar. Our own experience with the trocar in the horse is somewhat limited, but quite extensive in cattle practice, with never failing results.

THIRD CLASS.

The third class is essentially active and is manifest by violent pains, constant struggling, breathing short and quick, full, quick pulse, countenance expressive of great suffering, perspiration abundant in some parts of the body, the eyes have a wild staring look, the pulse becomes more quick and weak until finally they are imperceptible. To an experienced eye this diagnosis is not hard. The prognosis (termination) is always serious, although intestinal congestion may be treated with success if undertaken early. It has generally passed the curable stage before the practitioner is called.

TREATMENT.

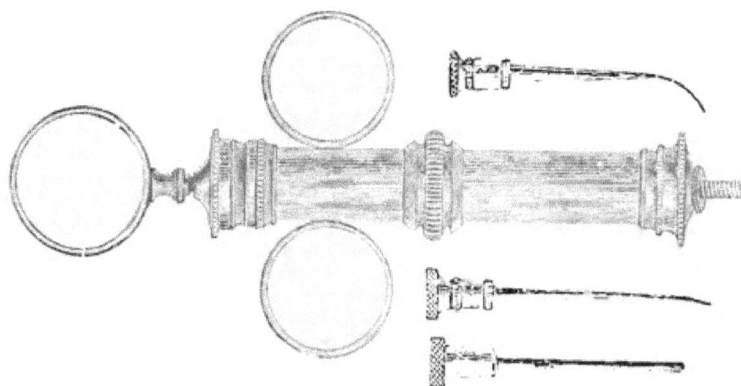

[THE HYPODERMIC SYRINGE.]

This class being essentially active it requires prompt and energetic treatment. First give with the hypodermic syringe* from four to six grains of morphine. Dissolve the morphine in one syringe of water and inject it under the skin of the neck. Then give of good whisky, one-half pint; tincture of capsicum, one ounce; tincture of ginger, two ounces; raw linseed oil, one half pint; mix. Give all at once and repeat your injections of morphine as often as may be necessary to

* If you have no Hypodermic Syringe give 8 to 15 grains of morphine in a little water.

keep the animal quiet. In one hour from the first dose give 30 to 40 drops of tincture of aconite in a little water; repeat this dose once every six hours until you have given from four to six doses. If the acute symptoms have not subsided by the time you have given the first dose of aconite you are to give following: Chloral hydrate, four drams; bromide of amonia, two drams; warm water, one-half pint. Mix and give all at once, and repeat this drench as often as may be necessary to keep the animal quiet.

[THE INJECTION FUNNEL.]

Give warm water injections per rectum often. No harm can come from injections even in large quantities. In the absence of a syringe you can take about two feet of rubber hose and a tin funnel, fix the funnel in one end of the rubber, then grease the other end and insert it into the rectum and pour the water in through the funnel. We generally use this in preference to the syringe.

FOURTH CLASS.

The fourth class includes colic caused by foreign bodies of different nature becoming lodged in the intestinal canal; a hard thing to diagnose and serious in its nature.

Give a physic ball or drench. (See Index). Then give the treatment recommended in the first class.

FIFTH CLASS.

Displacement of the bowels is generally discovered after death by post mortem. The diagnosis of some forms, however, is known to the skillful practitioner before death; such as scrotal Hernia (commonly called breached or ruptured in the

scrotum, etc.) This complication of intestinal trouble has
singled out some very fine stalli ns, among which we might
mention Duke of Perche, the property, at death, of Mr. E.
Wo.dman, of Paw Paw, Mich.; the Geronomus High stallion,
of Decatur, Mich., and many others have been reported to us
as having died from strangulated hernia (by the bowels passing down into the scrotum). It appears, from reports that
the average veterinary surgeon is unable, first, to diagnose,
and, secondly, is unable to remedy the wrong when once diagnosed. We beg leave to say that it is very easy to diagnose
and not so difficult to remedy when once you know how, at
least, we have been successful in every case operated upon in
the last 10 years (15 to 20 cases). We have seen some animals suffering from hernia (breached) seeking relief by bal-
ancing themselves on their back in the corner of the stall.
But this symptom is no positive proof, as many animals assume this attitude in knotting of the bowels (gut tie), rupture
of the bowels and obstructions. If strangulated hernia is not
attended to, early mortification will take place and death
ensue.

TREATMENT.

Whenever you see a stallion or gelding exhibiting sym-
toms of colic the first thing to be done is to examine the scrotum carefully well up in the groin, feeling for any unnatural
enlargement. Should you suspicion anything wrong you are
to roll up your shirt sleeve to the shoulder and grease the arm
with fresh lard, sweet oil or common soap, and carefully examine per rectum. Remove the excrements, then work your
hand gently into the intestine and down to the place where
you think the gut has passed out into the scrotum; now rub
carefully over this place and you can feel the imprisoned
bowell. If the rupture be on the left side of the horse, put

your right hand in the rectum and the left hand between his legs, and vice versa. Spread the fingers of your right hand around the imprisoned bowell, holding them quite firmly against the abdominal wall, now with the left hand force the bowel carefully up into the right hand. By careful manipulation you can usually reduce the hernia (rupture). Should you fail to return the bowel you will be compelled to send for a qualified veterinary surgeon at once, who will cast the animal and return the bowel by an operation.

MEDICINAL TREATMENT.

Take of raw linseed oil, eight ounces; sweet spirits niter, two ounces; mix. Give all at once followed by scalded bran mashes for a day or two.

SIXTH CLASS.

Rupture of the stomach or large intestines may be recognized by sudden relief or complete disappearance of the colic, while at the same time the general sickly condition of the animal gradually increases. The increase will be marked by the weakening of the pulse, the gradual cooling of the body, followed by a cold sweat.

Hence it must be acknowledged that there are genuine and serious difficulties in the way of the practitioner, yet, when in the presence of a suffering animal, if the practitioner will bear in mind the data that he must possess in his physiological knowledge and will rapidly and carefully analyze the symptoms of the case before him, comparing the positive and negative symptoms, weighing the case in his mind, comparing the acts in similar cases within his memory, the experienced and judicious veterinarian may reasonably hope to reach a

satisfactory and nearly accurate diagnosis. This is the most
important point to reach in order to establish a proper theory
and mode of treatment to discover and reach the cause, if
possible, and to remove it and save the life of the patient.
This is the work which the veterinarian must accomplish; this
is what he is for, and should he fail in one case, he is branded
as an ignoramus. It is all right for people to die, but horses
must not die. No, no.

However we are like the Irish M. D. Very successful
indade, sir. Niver lose a patient at all, at all, sir. No; not
one sir; except in their last sickness, and thin, somehow,
they'll die in spite of faith, sir, or midisin'."

CHRONIC IMPACTION OF THE COLON.

This disease is most prevalent in the early spring
months, and attacks horses that have been fed on dry, overripe
food, which is usually straw; an article unfit for food.

SYMPTOMS.

The animal may be found lying down, but will get up
when spoken to, but soon lies down again. He may continue
to lay around for several days without exhibiting much pain
until the acute symptoms set in, when he will roll and tumble
about for a while; then a deceptive improvement takes place,
he stands quiet for a while, when the head is thrown into the
air, and with a few convulsive efforts death ends his misery.
Again an animal may paw a little with one front foot or
stamp with the hind feet, lie down and remain perfectly quiet

for a while, then get up and eat a little, then lie down again. By lifting the abdomen the bowels seem heavy; you can also detect a fullness and hardness of the colon by oiling your hand and carefully introducing it into the rectum. Each day he seems a little more restless, getting up and down more often, and if nothing is done to relieve the symptoms the animal will die about the same as the former.

TREATMENT.

Take of powdered barbadoes aloes, twelve drams; powdered nuxvomica, one dram; powdered podophyllin, one dram; molasses or soap to make into a ball and give all at once, or you may dissolve the medicine in warm water and give as a drench. Then give warm water injections per rectum and freely, giving the following every six hours: Fluid extract of nuxvomica, one ounce; whisky, eight ounces. Mix and give one tablespoonful at a dose. If the bowels do not move in 36 hours take of linseed oil, eight ounces; sweet spirits of nitre, two ounces. Mix and give all at once. Continue the whisky and nuxvomica as before, and repeat the oil once every 24 hours until relieved. Should the animal exhibit much distress you had better take of chloral hydrate, four drams; hyposulphite of soda, two ounces; water one-half pint; mix. Give all at once and repeat as often as may be necessary to keep the animal quiet. Termination usually favorable.

CHRONIC IMPACTION OF THE RUMEN.

MAW-BOUND, DRY MURREN, HOLLOW-HORN AND WOLF IN THE TAIL

Are names by which we have heard stock growers speak of this disease.

The stomach becomes filled with dry food and the muscular coat is paralyzed from the excessive weight.   An unusual amount of corn, oats, bran or middlings is liable to produce it.

SYMPTOMS.

The nose is dry, ruminating stopped, the animal is dull and seems to suffer some pain which is indicated by a grunt or groan when breathing, more or less restless.   There is a heaviness and fullness of the abdomen.   By pressing on the left side the rumen (stomach) has a doughy feeling, and you can leave the indenture of your hand on the stomach, the feces are covered with slime.   Food may remain in the stomach of a cow for a long time without causing death.

TREATMENT.

Take of epsom salts, sixteen ounces; croton oil, one dram; powdered aloes, eight drams; boiling water enough to dissolve.   When cool give all as a drench.   Twelve hours later you will take of raw linseed oil, one pint; good whisky, one-half pint; sulphate quinine one-half dram.   Mix and give all at once, and repeat this every 12 hours until the bowels move freely.   In all bad cases it is a good practice to give injections per rectum of warm water every two or three hours. Occasionally which is the exception and not the rule, we find a case that resists all medicinal treatment.   Then we have to resort to *Rumen Otomy*.   (See Index.)

TYMPHANITES IN THE COW.

HOVEN, CLOVER BLOAT, BLOATING, ACUTE INDIGESTION, ETC.

This disease at times kills very quickly and at all times is to be treated energetically.

Causes.—In some localities clover, while wet with the

dew, will produce it very quickly. Sudden change of food or an over feed of rich food such as chop or corn meal; in fact any kind of food may produce it. It may be a symptom of choak or other disease.

SYMPTOMS.

The left flank is swollen to a great extent and there may be belching of wind or gases. Ruminating (chewing the cud) is suspended. The rumen (paunch) sounds hollow when you tap it with your hand, the head extended and the mouth may open sufficiently to allow the tongue to hang out. Toward the last the eyes are bloodshot, the animal staggers as it moves and if relief is not given at once soon dies from asphyxia.

TREATMENT.

[THE TROCAR AND CANULA USED IN TAPPING CATTLE.]

The first thing to be done in bad cases is to make an opening in the rumen (stomach) at once, which is best done with the trocar and canula, and is done as follows: Take a sharp knife and cut a hole just through the skin a half inch long mid-way between the point of the hip and last long rib on the left side and about one inch below the point of the hip. Now take the trocar and canula and force it through the opening you have made in the skin into the rumen, incline the trocar, with the point a little forward introduce it the full length. Now withdraw the trocar leaving the canula in; this forms an opening through which the gas escapes. As soon as

the bloat has gone down cork up the canula and allow it to
remain for a while removing the cork as the gas accumulates.
As soon as you have relieved the urgent symptoms you are to
take of carbonate of amonia, six drams; spirits of turpentine,
two ounces; raw linseed oil, one quart; mix.   Give all at once
as a drench.   If not relieved in one hour take of epsom salts,
eight ounces; quiniue, one dram; bicarbonate of soda, two
ounces; boiling water enough to dissolve   Cool, shake well
and give as a drench.   As soon as the animal quits bloating
remove the canula and bathe the wound with alcohol, arnica,
camphor, whisky or cold water, and give the following for a
week: Fluid extract of nuxvomica, two ounces; water, one
pint.   Mix and give one tablespoonful three times a day in
the feed or on the tongue.   The termination of this disease is
usually favorable.

CATARRH IN THE COW.

HOLLOW HORN, WOOLF IN THE TAIL.

This is a disease of the air passages and differs somewhat
from catarrh in the horse and the general symptoms are a discharge from the nose, does not sweat as in health, but is dry
and rough, there is fever and a cough present, pulse a little
quick, the animal does not ruminate (chew the cud), the extremities are cold and the disease sometimes becomes so aggravated that there is pus or matter in the cavity of the horns.
This gives rise to an imaginary disease called *hollow horn*.
There is no such disease.   All cattle have hollow horns—if
they have horns—a fact which all stock growers ought to
know.   Another symptom of catarrh is a dropping apart of
the bones at the end of the tail.   This is due to a relaxed con-

dition caused by disease, and gives rise to another imaginary ail called *woolf in the tail.* As soon as the animal is restored to health the tail regains its natural condition, and we hope that the "Young America" of to-day will discard these erroneous ideas, get out of these old ruts that our ancestors have wallowed in for so many generations, and post themselves before they make an assertion or go to dosing a poor dumb animal for some imaginary disease.

TREATMENT.

Give from one-quarter to one-half pound of epsom salts dissolved in warm water as a drench, then give powdered saltpetre; one and one-half tablespoonfuls three times a day for a week. Then give powdered sulphate of iron, one tablespoonful three times a day for three or four days. Feed on good, laxative food.

LARYNGITIS AND PHARYNGITIS.

AN INFLAMMATION OF THE THROAT.

This disease is characterized by loud breathing, swelling of the throat, ruminating (chewing the cud) suspended, much difficulty manifest while drinking.

TREATMENT.

Apply the mercurial blister to the throat, neck and under jaw, extending well up to the ears, then give raw oil, one pint; spirits turpentine, two ounces. Mix and give all at once as a drench. Then take powdered chlorate of potash,

one pound and give one tablespoonful every four hours on the
tongue. Feed on sloppy food, blanket the body and attend to
the general comfort of the animal.

RUMINATING.

THE COW'S CUD.

All animals that rechew their food are called *ruminants.*
This class includes the cow, sheep, goat, etc. The stomach of
these animals has four compartments, the rumen, the reticulum, omasum and abomasum. The rumen is the large part of
the stomach which the food enters when swallowed. After
the cow has finished her meal she will remain quiet and remas-
ticate or rechew her food which comes up into the mouth in small
quantities, is rechewed and again swallowed when it passes on
into the reticulum, omasum (manifold) and so on through.
Another quantity is forced up into the mouth, chewed and returned in a like manner until all has been rechewed. This
is called *ruminating* (chewing the cud). This being a fact,
how absurd it is to talk of a cow

LOSING THE CUD.

Yet there are many people who believe that a cow has a
permanent cud of some kind although they can not tell you
what it is like, but they are sure she loses it occasionally and
cannot find it. So they wad up a dish rag, a bunch of hay or
grass, a large piece of fat pork, or catch a live bull-frog and
force it down her defenceless throat for the purpose of making
a new cud. How utterly ridiculous the idea. Too ridiculous
to talk about. Whenever the cow eats she has a cud, and

will chew it if she is able to. If she does not chew her cud it is because she is sick, in which case you should endeavor to find out the ail and give the proper remedies, which will restore her to health, when she will find her cud without your assistance, and chew it as vigorously as the young lady of to-day does her gum.

AZOTURIA.

STOPPAGE OF WATER, PARALYSIS—KIDNEY DISEASES, ETC.

This is a disease peculiar to fat, easy keeping horses, those which take on flesh easily.

Causes.—Allowing an animal to remain in the stable for a few days without exercise, continuing the usual work diet, the building up process goes on faster than the excretory organs can carry out the worn out matter. Consequently the urea (poison part of urine) and hypuric acid on exercise, are re-absorbed into the system, acting as an irritant, causing spasms and contractions, especially of the large muscles of loins and hip. The faster the work the more severe are the symptoms and serious the attack.

Prof. Smith, of the Toronto Veterinary College, says that azoturia is more common in the winter months, but my experience has led me to the conclusion that it is most prevalent in the United States in the early spring months, when farmers commence to feed up their horses for spring plowing, which they have commenced, perhaps, when a rain storm sets in for two or three days and it takes (in clay soil) two or three days for the ground to dry, during which time the horses must be fed high (says the farmer) that they may be ready for the extra work caused by delay on account of the rain.

SYMPTOMS.

The animal (always a good one) when taken from the stall appears in the best of condition. This is a common expression "he never felt better" he is driven from a few rods to a few miles when he begins to sweat profusely, then he begins to lag and if stopped you will notice a nervous twitch of the muscles, trembling of the flank and sometimes of the shoulders, these symptoms are speedily succeeded by a loss of motor power, and a desire to lie down, or may go lame *suddenly as though he had picked up a nail, or fall down suddenly. Sometimes they are very restless, trying to get up, tumbling around bruising themselves badly. The bowels are costive, the urine is of a dark coffee color. From what has been said about azoturia it will be readily seen that a horse may be only slightly or severely attacked.

TREATMENT.

If the animal is very restless give at once a hypodermic injection of morphine, six grains; if you have no syringe give instead, chloral hydrate, four drams; water, one-half pint; dissolve and give all at once as a drench. If quiet, this dose is to be omitted, but the following must be given without delay, regardless of the severity of the symptoms. Take of powdered

* NOTE—The lameness in azoturia differs from all other forms of lameness; but slightly resembles a broken leg as it forms no column of support whatever, it is sometimes mistaken for stifle joint affection from which it differs materially, in stifle affection the leg forms a column of support when compelled to move the three unaffected ones. In azoturia the reverse is the case, the affected leg forms very little or no support, the animal moving as though he had but three legs.

barbadoes aloes, ten drams; powdered nuxvomica, one dram; powdered capsicum, one dram; powdered ginger, two drams; molasses or soft soap enough to make a stiff mass, make into two pills and give both at once, or you may add one-half pint of hot water and give as a drench. Give warm water injections per rectum every two hours until the bowels move freely. If restless give bromide of potash, but if quiet give nitrate of potash in tablespoon doses every six hours, alternate with the following, fluid extract nuxvomica, one ounce; alcohol, eight ounces; mix, give a tablespoonful every six hours, this brings one dose every three hours, turn about. If the bowels do not operate *freely* in twenty-four hours, take of raw linseed oil, twelve ounces; spirits of turpentine, two ounces, mix, give all at once and repeat the oil alone every twelve hours until the bowels operate freely. Should the animal be unable to stand, make a good soft bed and turn him over about every three hours. As soon as you have given the aloes prepare the following, take of spirits of turpentine, aqua ammonia, sweet oil and alchohol, each two ounces; mix, shake and bathe the loins, hips and legs, then blanket and keep him warm. Some recommend the applications of blankets rung out of hot water and applied to the loins and hips. This is good or the best treatment if you will only continue it long enough, say twelve hours at a time or longer. For the medicinal effect of water *hot* or *cold* is proportionate to the length of time applied. Do not be in a hurry about getting the animal up until the third or fourth day. About this time put the sling under the animal and raise him upon his feet, have plenty help and as soon as you get him up bathe his legs with the liniment and have them hand rubbed for ten minutes vigorously, after which he will usually try to stand but should he fail to regain the use of his limbs after a

[THE SLING.]

few minutes you must let him down renewing the attempt once
a day until he does stand.　The urine (water) should be drawn
at once and should the animal fail to urinate it must be drawn
every six hours.

SHARP & SMITH CHICAGO

[THE FEMALE CATHETER.]

The male catheter is three feet long, made of rubber.

PURPURA HÆMORRHAGICA.

An eruptive, non-contagious fever of an intermittent type,
usually occuring as a sequel to other diseases, such as influenza

and its origin. Putting an animal to work too soon after recovering from any debilitating disease is a common cause of
purpury. Damp ill-ventilated stables where the animal is
compelled to inhale the products of decomposing urine and
manure is another cause.

SYMPTOMS.

The first appearance is a swelling of the legs, it may be
slight, at the hocks only, or the whole leg may be swollen to
the body, usually all of the legs are swollen and at times so
large that the animal cannot lift them from the floor. I have
seen a few cases that the swelling stopped so abrupt at the
body that it looked as though a string had been tied around
to prevent it from going further. I have treated some cases
where the swelling of the legs would begin to go down on the
fourth or fifth day and on the evening of the fifth day the legs
were natural, but the next morning I found an enlargement at
the lower part of the neck or upper part of the breast the size
of a child's head, hot and tense this continued to increase in
size for four days when I opened it, and I think that there was
at least six quarts of thin white pus escaped and the animal
made a good recovery. Another case similar to this except
the swelling appeared in the fleshy part of the hip. The pulse
and temperature are changeable, in the morning you may find
the pulse 60°, the temperature 102°, and at night you will
find the pulse 80°, temperature 104½°. The animal usually
stands but occasionally will lie down and the chances are that
you will have to help him up, but do not disturb him for a
few hours if he lay quiet as the rest so obtained seems to refresh him wonderfully. This disease has been dubbed "YELLOW
WATER" by some on account of a yellowish looking fluid which

exudes (seips) through the pours of the skin, at times it
streams down the legs in pools where the animal stand. Occasionally but not often this fluid resembles blood, the bowels
are constipated and urine scanty.

The duration of this disease is from fifteen to fifty or
sixty days, the average time being about forty days. Remove
the animal to a well ventilated, dry loose box stall and give
the following laxative: Raw linseed oil, one pint; spirits of
turpentine, two ounces; mix, give all at once as a drench, repeat the oil as often as may be necessary to keep the bowels
regular. Take *powdered chlorate of potash*, sixteen ounces,
give one tablespoonful three times a day for three days, then
omit and take of *powdered sulphate of iron*, eight ounces, give
two teaspoonful three times a day for two days, then return to
the potash for three days and then back to the iron and so on.
Feed on laxative food, scalded bran and oats, allow plenty of
water, milk, eggs, etc. If the legs crack, ulcers form or the
flesh sluffs in spots, bathe with "WHITE LOTION", (See Index.)
As soon as he begins to move around allow him a yard to run
in. Though the animal is often a pitiful looking sight, the
termination of the disease is usually favorable.

CHRONIC COUGH.

This may be the result of either throat or lung trouble,
and must be treated according to what you think the true
cause.

If from throat troubles apply a fly blister (See Index) to
the throat extending down between the angles of the jaw and

well up toward the ears. Repeat the blister once a month until cured. Take of iodide of potash, two drams; powdered muriate of ammonia, two drams. Mix and give as one dose; repeat night and morning for a week.

Another good prescription is camphor, digitalis, opium, and calomel, of each one dram. Make into a ball and give one at night for a week. omitting the calomel after the second night. It is a good practice to give the *turpentine drench* the first thing, followed by a bran mash and the cough remedy.

SCRATCHES.

The most common cause of this disease is allowing the legs to dry by evaporation. This sets up an irritation. Heavy horses are more liable to it than the better bred lighter ones are. The hind legs are more often affected than the front ones.

SYMPTOMS.

The horse may be stiff and sore on coming from the stable, but gets better on exercising. The animal may lift its leg as in string-halt, the fetlock somewhat swollen and may bleed. Scratches may terminate in greese.

TREATMENT.

Give a physic ball (See Index), first giving a scalded bran mash with plenty of salt the night before and little or no hay. Give all the water the animal will drink, continue the bran mash for three or four days, giving powdered nitrate of potash in tablespoonful doses three times a day for a week. *Externally* use "*white lotion*," carbolized cosmoline, or the "*white healing powder*," washing the parts as often as you may deem necessary and remember to always rub the legs dry and not allow them to dry in the air, as this is a sufficient cause to produce scratches.

GREESE.

This might be be said to be a more advanced stage of *Scratches*, and should be treated the same, although it is more difficult to treat and the results are not so favorable in all cases.

MUD FEVER.

This you might call another and more severe form of greese, in which the hair comes off of the legs, and is to be treated about the same. You will give internally, in addition to the above treatment, one dram of iodide of potash three times a day for a week or two.

DIABETES.

This disease is characterized by great thirst, an excessive discharge of urine, debility and rapid wasting away. I have known fat, nice horses to be attacked and in six weeks were mere skeletons. The causes are attributed to over-ripe, non-nutritious food, musty hay, oats or corn are the exciting causes.

SYMPTOMS.

The animal has an excessive thirst and profuse urination, rapid wasting of flesh, the lining of the nose and tongue often present a pale or sometimes a rusty appearance, the appetite is not much affected in the horse, but the dog has a ravenous appetite for flesh, the pulse is usually slower than natural, the breath sometimes smells sour.

TREATMENT.

The first dose to be given is crystal iodine, one and one-half drams; iodide of potash, one dram; water, two ounces. Mix and shake until dissolved, then add one-half pint of raw oil and give all at once as a drench. Wait 12 hours and give the following: Crystal iodine, three drams; iodide of potash, two drams; water, eight ounces; mix. This makes three doses (to be given in eight ounces of raw oil). Give one dose every 12 hours until all three are given. Then take of iodide of potash, four ounces; crystal iodine, four drams; water, sixteen ounces. Mix and give one tablespoonful in the feed three times a day. One of the essential things is a complete change of food, regardless of what they have been eating.

ACUTE LAMINITIS OR FOUNDER.

This is one of the most dreadful diseases that horse flesh is heir to. You cannot imagine how the poor animal suffers when once he is attacked with acute laminitis.

Causes.—It is caused by drinking cold water when heated, cooling off too suddenly, standing in a darft, a sudden chill, an abortion, being compelled to stand in a constrained position, over-work, eating highly nutritious food, such as wheat, rye, etc. Inflammation of the bowels and bronchitis are also liable to terminate in laminitis; driving through a stream of cold water when the horse is warm, standing in a stall on a heap of hot, decaying manure is another fruitful cause.

SYMPTOMS.

When only the front feet are affected the animal will stand with the hind legs well under the body and the front legs somewhat extended and moves with great difficulty. In examining a suspicious case force the animal back ; if it be founder he will roll back on his heels, raising the toe from the ground. When the hind feet alone are affected he will stand with all of his feet drawn together, a picture of distress. Usually he will not stand long at a time when the hind feet are affected, but will lie down, and seems to get immediate relief as is indicated by the pulse falling 10 to 20 beats in the course of as many minutes. When one foot is alone affected as it sometimes happens when a horse for some reason is compelled to stand on one foot for a long time. For example a horse has a nail or gravel in the right hind foot; he stands on the left one for some time. On your next visit you find him standing on the lame right leg with what was the well foot extended under the body with the heel resting on the floor. If but one front foot is affected it will likewise be extended. When all four feet are affected you will have a combination of the foregoing symptoms with heat in all the feet. and the legs may be somewhat swollen, which is a good symptom ; the breathing is quickened, the pulse are full and quick, 60 to 80 beats per minute.

TREATMENT.

Give a laxative ball or drench at once, then give the following alternately (turn about) every hour : Fluid extract

aconite root, one ounce; water, one pint; mix. Give one
tablespoonful every two hours on the tongue. Powdered
nitrate potash, eight ounces. Give one tablespoonful every
two hours on the tongue. As soon as the appetite returns
give scalded bran mash and remember that all diseases requir-
ing a physic requires water also; a little at a time and often.
As soon as you have given the first dose of medicine remove
all the shoes and rasp off the wall of the diseased feet so that
the sole will come to the ground, then put on an "oil cake meal
poultice" (See Index). Be sure to have the poultice quite
warm and pour warm water over the poultice every two or three
hours. Renew the poultice once a day. Continue the aconite
and potash until the animal begins to improve, then gradually
diminish it in quantity and frequency. If the bowels do not
move pretty freely in 24 hours after the physic is given you
must give from one-half to one pint of raw linseed oil (according to the size of the horse) once a day until they do
more freely. About the third or fourth day you must commence
exercising the animal by leading or driving it a few rods and
back three times a day gradually increasing the distance up
to an hour at each walk. Should any lameness or stiffness
threaten to remain a mild blister is to be applied around the
top of the feet. In addition to this treatment I usually take
from two to four quarts of blood from the jugular vein the
very first thing we do in all bad cases. In the milder ones
there is no need of bleeding, but I do think it a good practice
to bleed in severe cases. We have treated a great many cases,
all of which have terminated favorably, and are sure that if
this treatment is carefully followed out recovery will be sure
and speedy, many cases recovering in from three to ten days·

BOTS IN THE HORSE.

3

1

[NO. 1, GAD FLY.]
[NO. 3, BOT.]

There is a great diversity of opinion with regard to the good and bad effect of bots. Some say that a horse cannot live without them, and assign as a reason, first, that the bots are loose in the stomach of a healthy animal and assist in the digestion of the food in a like manner and are as essential as the gravel in the chicken's gizzard. This is impossible, from the fact that the action of the stomach and bowels are such as would expel them at once. Besides, during their stay in the digestive organs of the horse they have no means of locomotion. This being a fact how absurd it is to think that the bot crawls up into the animal's throat and chokes him or brings about disease as is claimed by some. Others claim that bots have been known to eat holes through the walls of the stomach. This is utterly impossible, as the bot has no teeth, cannot bite, and does not live by eating, but receives its sustenance by absorption. That bots are capable of injury and occasionally are the cause of very grave lesions cannot be denied, but these injuries are rarely, if ever, properly diagnosed until after death. Post mortem has revealed to me the following: First, that the head or absorbent end of the bot is so deeply inserted into the mucous membrane of the stomach that no medicine will affect them except it be strong enough to destroy the horse. Second, that a few bots do no harm while in the stomach, and many only impair digestion, the cause of which can only be guessed at; third, that the only danger to be apprehended from bots is during their transit through the alimentary canal by attaching themselves to the walls of the intestines, forming a mechanical stoppage of the passage, causing rupture of the bowel. For this reason, if nothing else, we would advise every horse owner to keep the larvæ (nits) off of the hair, which is easily done by scraping with a knife blade. The many horses reported as dying from bots eating through the stomach were in all probability cases of gastritis, causing rupture of the stomach, allowing the bot to pass into the abdominal cavity, the "hoss" doctor branding the innocent little bot as a murderer in the first degree.

THE HISTORY OF THE BOT.

During the latter part of the summer the common gad fly attacks the horse, its object being not to derive sustenance, but to deposit its eggs on the hair, and it does this by means of a gluey substance which causes the ova, or egg; to adhere to the hair. The gad fly very wisely deposits its eggs on the shoulder and fore legs of the animal that it may be in easy proximity to the mouth. The animal heat hatches the eggs, which bring forth a small worm, the slightest motion of which causes the animal to bite himself, when the larvæ (as they are now termed(are taken into the mouth and along with the food and drink are conveyed into the stomach. Of course a great many of the larvæ are destroyed during their transit from the hair to the stomach. Some are dropped from the mouth, while others are crushed during mastication. Notwithstanding the great waste the interior of the stomach may become nearly covered with the larvæ, which are retained by means of two small hooks by which they attach themselves to the mucous membrane where they remain, sucking the juices of the stomach, until they are full grown, covering a period of about eight months, at the expiration of which time they voluntarily

let go and allow themselves to be carried along the alimentary
canal until at length they are expelled.   During this transit it
is again supposed that many more lives are lost, but those falling on suitable soil are warmed by the sun for a few weeks
until a third change takes place and the gad fly (or nit-sticker,
as they are sometimes called) is hatched out and goes about
the work of reproduction.   Thus it will be seen that the gad
fly uses the stomach of the horse as a medium of propagation,
perpetuating its kind.   And a strange kind it is, too; having
no mouth, it can take no nourishment, consequently its life is
short but fruitful.

NEMITODA.

HUSK IN SHEEP—LUNG WORMS IN LAMBS AND SHEEP.

These are round, thread-like worms from two to four
inches long, and are, perhaps the most important class of
worms as they have a head capable of penetrating any part of
the animal body except bone.   I have seen large flocks destroyed by the lung worms.

SYMPTOMS.

We notice a flock of lambs coughing.   They are out of
health.   The cough is paroxysmal and threads of mucous are
coughed up, and if we examine this mucous we will likely find
one of the minute thread-like worms.   This is a positive symptom.   On post mortem I have found the bronchial tubes
completely filled with worms.   Mr. Geo. Pearl, of Riverside,
Mich., lost about one hundred head of sheep during the winter

of 1888-89 from lung worms. These sheep wasted away to mere skeletons before death.　•

TREATMENT.

Change pasture, then place the sheep in a close building where they will be compelled to inhale the fumes of sulphur, which is to be burned quite freely once a day, giving internally one tablespoonful of turpentine in two ounces of raw oil every other day for a week, then every three days for a while. In very bad cases I would recommend the injection of one-half teaspoon of turpentine into the windpipe with the hypodermic syringe (See Index). You will sometime find these worms in calves (called hoose) and pigs, which are to be treated the same as sheep.

WORMS IN HORSES.

Horses are affected with worms and bots. The term hoose in cattle and husk in sheep are due to the presence of small worms in the lungs. Worms in the horse are of different varieties and are often a source of great irritation.

SYMPTOMS.

The coat staring and unthrifty, ravenous appetite, but the food taken apparently does no good. Occasionally an animal may be infested with worms in numbers sufficient to cause dropsical swellings of the legs, which may extend along the belly. The positive symptoms are the passage of worms in the feces, or imprisoned in the anas. Very often you will see a light-colored, floury substance around the anas caused by the worms having been imprisoned. The motion of the worm causes the animal to rub his tail against any convenient object.

TREATMENT.

Give a scalded bran mash well salted at night. The next morning give a *purgative ball*. In 36 hours give a turpentine drench followed by powdered copperas in a teaspoonful dose three times a day for a week, or until the feces (manure) is black, then give a *laxative ball* followed by the turpentine drench every 24 hours until the bowels move freely. Then give quinine in 15 grain doses three times a day until the appetite is good and the animal begins to thrive.

SCOURING ON THE ROAD.

This may be be due to bad or defective teeth, dyspepsia, or a relaxed condition of the bowels from any cause. These animals are sometimes called " WASHY " HORSES.

SYMPTOMS.

The animal when taken from the stable appears to be in good condition, but after you have driven a short distance the bowels begin to move off quite freely, and each stool is a little more moist and watery. Nervous, bad, or long coupled horses are the ones most often affected.

TREATMENT.

First examine the mouth, and should you find anything wrong with the teeth correct it, and then look to the food, as it should be of the best quality. If the quality is all right then look to the quantity, which must be regulated. Internally give the turpentine drench followed by powdered sulphate of iron in teaspoonful doses three times a day for a week. As a preventative give such horses one ounce of tincture of opium in a gill of water on leaving the stable for a drive.

DIARRHŒA—HORSES.

Narrow-loined, flat-sided, spindle-shank, long-backed horses are the ones most often affected, although a well-built, nervous horse may be affected.

TREATMENT.

Raw linseed oil; eight ounces; tincture of opium, two ounces; good gin, four ounces; mix. Give all at once as a drench; followed by powdered sulphate of iron in teaspoonful doses three times a day for a week.

Should the diarrhœa continue after 12 hours take of tincture of opium, two ounces; prepared chalk, two ounces; raw flour gruel, one pint; mix. Give all at once as a drench and repeat the dose every four hours until the purging is checked.

DIARRHŒA—CATTLE.

DYSENTARY, BLOODY MURREN, BLOODY FLUX.

This disease is seen in both the acute and chronic forms in cattle. It may be induced by bad food or putrid water.

SYMPTOMS.

In the acute form there will be shivering fits, arched back, the animal grunts at every breath, grinds its teeth, a frequent discharge from the bowels of a muco-purulent matter mixed with small lumps of feces and blood, there is much straining, some abdominal pain evinced by the arched back and whisking of the tail, great dullness and rapid wasting away. In the *sub-acute* or *chronic* form we have the foregoing with the additional symptoms, looseness of the teeth. the feces is discharged involuntarily, the eyes dim and sunk into the head, the feces contains a gaseous material, which causes the appearances of air bubbles upon their surface when expelled from the body.

TREATMENT.

Place the animal in a good dry place with plenty of good dry straw for a bed. Then take of raw linseed oil, two pints; tincture of opium, two ounces; spirits turpentine, two ounces; tanic acid, two drams; mix. Give all at once as a drench. Six hours later take of quinine, one-half dram; tincture of opium, two ounces; raw flour gruel, two pints. Mix and give. Repeat this dose every six hours until relieved. Give rectal injections of raw oil, one-half pint, every time you give the medicine. As soon as the diarrhœa is checked and the animal begins to improve give sulphate of iron, one-half tablespoonful three times a day in feed or on the tongue for a week or so.

ANASARCA.

SWELLED OR STALKED LEGS, SWELLED SHEATH, ETC.

This is due to a relaxed condition of the capilliary system, the blood vessels become dilated, the walls thinned and the blood passes through into the surrounding flesh. This is called extravasation of blood into the surrounding tissue without which there is no *swelling*. Heavy horses and those with round fleshy legs are most liable to it. Horses stabled in idleness until they are out of condition, and then put to work, are liable to this ail. Washing the legs and allowing them to dry in the air is another cause. Scratches and cracked heels will produce it, or it may be the result of improper bandaging. Kicks, bruises or injuries of any kind will cause a swelling. If due to inflammation there will be pain; if not, there will be none.

TREATMENT.

Give a scalded bran mash, followed in 12 hours by a physic ball. Then give in mild cases the following: Powdered iron sulphate, four ounces; potash, nitrate, eight ounces; mix. Give one table spoonful three times a day in the feed or on the tongue until all is given.

In obstinate cases you will give the *alterative tonic* instead of the above prescription. *Externally* bathe the legs or swollen parts with the *white lotion* three or four times a day.

GLANDERS.

This is is the most loathsome of all diseases known to the equine race, and in preparing this description it has been my intention to give you the prominent diagnostic symptoms in plain terms, along with which I have given you what seems to me to be the birthplace of glanders—its first source or origin.

ACUTE GLANDERS

Is transmissable to all domesticated animals and man except the cow. The first symptoms are a rigor (chill) of the most persisting character, a rise of temperature to 105° or 106°, and it may run as high as 109°; the respiration hurried, flanks tucked up, the pulse feeble and rapid, the heart palpitating, the appetite gone, the visible mucous membrane at first of a dark, copper color with patches of ecchymosis of a dark red hue; these patches are speedily converted into ragged edged ulcers from which issues a copious discharge. The submaxillary glands enlarge and other lymphatic glands become inflamed, raise up suppurate and burst, discharging a purulent puss. The discharge from the nose may be from both nostrils, but usually from the left one only. The discharge is at first watery, then purulent. The peculiarity of the glander discharge is that of its being sticky, adhering to the nostrils, having a tendency to close them up. There may be fœter (stink), but it does not smell so bad as the discharge from nasal gleet, with which it is often confounded. The enlargement in the angle of the under jaw in glanders differs from other throat affections; the enlargment being on either side close to or adhered to the jaw bone. If the discharge is from but one nostril the enlarged gland will be on the same side as that of the discharge. Another peculiarity is that the discharge has a greenish cast. We also have other catarrhal affections, the discharge of which has a greenish cast at certain seasons of the year, but they are easily distinguished from glanders. I have but little doubt that some of my readers will ask the question: Where does glanders come from, or from whence is its first origin? Does it not come from horse distemper? No, sir; it does not! At least I am sure that horse distemper would have to degenerate into something else before it would run into glanders. Besides, I have never known or heard of such a termination. I will tell you the diseases most likely to degenerate into glanders—diabetes insipidus and its complications. Those diseases do, occasionally, degenerate into glanders; and some authors say that any debilitating disease is liable to terminate in glanders. This statement, I think, is untrue.

Again, some of you may ask, What is diabetes? Diabetes is a disease brought about by giving food which contains an undue amount of heating properties, such as musty hay or oats, which stimulates the urinary organs, increasing their functions; other organs become involved, the animal rapidly wastes away, changing from a fat, sleek horse to a mere skeleton in a few weeks. This great waste of flesh is the result of non-assimilation of food; that is to say that the food taken does no good; the building up process is, in some cases, entirely suspended, while the tearing down or wearing out process continues to increase daily, causing great waste and poverty of blood. This disease, I think, is the birth place of glanders.

CHRONIC GLANDERS.

This form of glanders is the one to be most dreaded, from the fact that there are instances reported where it could not be detected before death even by experts. And I know of no better illustration of the occult nature of this most loathsome disease than the one furnished on Page 324 of the State Board of Health for 1879, by Professor Spinola, of Berlin, Germany.

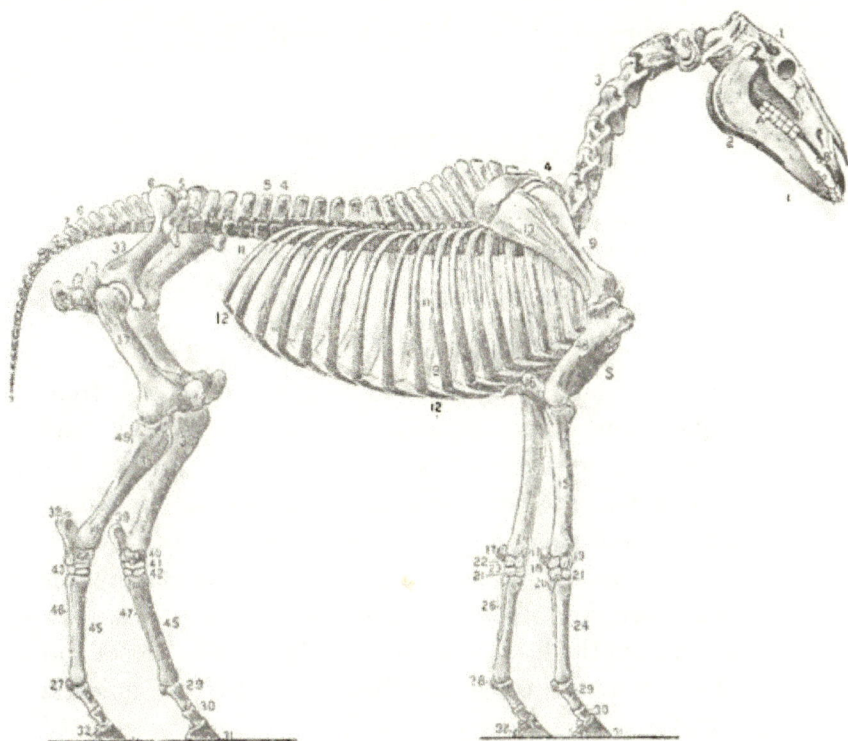

DESCRIPTION OF THE SKELETON OF THE HORSE.

1. Cranium.	15. Radius.	31. Ospedis.
2. Lower Jaw.	16. Elbow.	32. Os Naviculare.
3. Cervical Vertebræ.	17. Trapezium.	33. Pelvis.
4. Dorsal Vertebræ.	18. Cuneiform.	37. Femur.
5. Lumbar Vertebræ.	19. Lunar.	38. Tibia.
6. Sacrum.	20. Tropezoid.	39. Os Calsis.
7. Coccygeal Vertebræ.	21. Os Magnum.	40. Astralagus.
9–9. True Ribs.	22. Scaphoid.	41. Cuniform Magnum.
10–10. Cartilages of Ribs.	23. Unciform.	42. Cuneiform Meidum.
11–11. False Ribs.	24. Great Metacarpal.	43. Cuboid.
12–12. Cartilages of False	26. Small Metacarpal.	45. Great Metatarsal.
Ribs.	28. Sesamoid.	46. Small Metatarsal.
13. Scapula.	29. Os Suffraginis.	49. Fibula.
48. Humerus.	30. Os Corona.	S. Sternum.

LAMENESS AND SHOEING.

—

CHAPTER II.

Causes, Symptoms and Treatment, With Practical
Hints--Illustrated.

LAMENESS.

The word *lame,* according to some authors, comes from
the Anglo-Saxon word *lam,* weak; the term *lame* and *weak*
are synonymous in some parts of the country; thus it is a
common expression to hear "that is a lame excuse" for a poor
excuse, a "lame sermon" for a poor, weak sermon, etc.

Lameness is not a disease, but is a sign of disease; or
rather an expression of pain. Lameness may arise from diseases apart from the limb, such as spinal or nerve diseases and
occasionally from disease of the liver or kidneys. Lameness
may exist for a short time without disease; the expression of
pain from a gravel lodged in the' shoe, or an ill-fitting shoe,
but if the cause of pain be allowed to remain for any length
of time inflammation is sure to follow. Thus it will be seen
that disease exists much oftener without lameness than lameness without disease.

LOCATING LAMENESS.

The readiness with which some men are able to detect and
locate lameness seems to be an instinct or natural gift, while

to others it is no easy task. When a lame animal is presented for examination the first thing to ascertain which is the lame leg or legs. This may seem an easy matter, yet at times is attended with much difficulty. For example, a horse lame in the left front leg is trotted from you, he may seem lame in the right hind leg, for the quarters seem to rise and fall. But when the horse is trotted towards you it will be readily seen that the motion of the hind quarters depends upon the raising and dropping of the head and body. You will then see that the lameness is in the front leg. Another difficulty is the location; when the lameness is in both front legs such horses, although the head does rise and fall in proportion to the lameness, yet they have a rolling motion to the body quite perceptible to the close observer. Lameness in both hind legs is more easily distinguished. Each particular lameness will be fully explained under their respective heads.

THE HORSE'S FOOT.

There is no subject that calls for more attention than the consideration of the horse's feet. At the present time so great is the ignorance among the owners of horses that most of the lameness arises from the mismanagement of this all-important part of the animal. An old adage is "no foot, no horse." This is only too true. Writers upon this subject are numerous and nearly all have run in the same channel, promulgating false ideas from one generation to another to the great destruction of the usefulness of the horse. The prejudices are so great and deeply rooted that it is dangerous for any one to try to teach a more rational doctrine, and undoubtedly some of the readers of this article will brand its author as a crank or something worse.

Nearly every one looks upon the foot as a very wonderful and complicated piece of mechanism, and do not stop to consider that no matter how complicated it may be within it is enclosed in a simple horny box, and that all efforts of shoeing should be directed to preserve that box in a natural condition and its position in relation to the limb should not be altered by shape or form of the shoe. Some believe that the horny foot is an elastic organ and that its elasticity should be kept intact by paring the sole, peculiar nailing on of the shoe by keeping the wall as moist as possible, etc.; while some claim that mechanical advantage can be given to the tendons by the form and weight of the shoe. All these ideas are errors and have proved themselves as such when put into practice.

The hoof is built of tubes matted together. These tubes are similar to hair and are secreted by the same kind of cells. Horn is spoken of by some authors as being built or hairs firmly matted together. *The Wall* of the foot is the part which is visible when the foot rests on the ground, and is divided into toe, quarters and heels. *The Sole* is a thick plate of horn which occupies the ground surface of the foot except that part which is protected by the frog and bars. *The Frog* is a prominent mass of spongy horn lodged between the bars, filling up the triangular space. The horn of the frog differs considerable from that of the wall or sole, being much finer and softer. The color of the hoof varies, but usually corresponds to that of the neighboring skin. The hoof, which forms a horny box, contains the pedis bone, the navicular bone and the coronea bone, the lateral cartilage (soft part of the heel), the sensitive frog, the sensitive lamina and sensitive sole, the ligaments and tendons which forms the attachments, the blood vessels and arteries which supply life and nourishment; also the nerves which impart the sense of feeling. The shape of the horse's foot is by many said to be round—describing a circle. While this may be true of some feet it is not true of all.

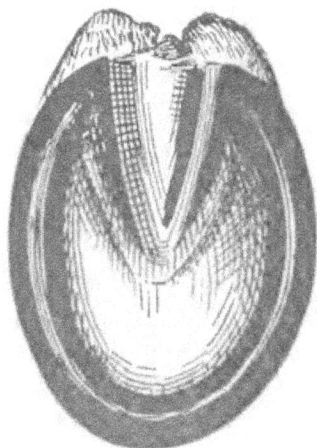

[THE RIGHT FOOT.]

Being a practical as well as a theoretical shoer, I am able to speak from experience. I have found many, in fact the majority, of all feet with the inside of the hoof less rounding than the outside and the longest point of the foot inside of the center at the toe, as is shown in the cut. This prominence or long point the average horse shoer removes by making the shoe to fit his eye (round), which he nails on the foot; then he takes his toe knife and chops off the inner part of the foot which so changes the swing of the foot that it comes in contact with the fetlock of the grounded foot, causing what is called *Interfering*, of which we will speak later on.

In the first place, taking the normal foot as a guide, the frog, bars and sole of the foot should never be cut away, as is the common practice. Secondly, all calkings and toe pieces should be done away with for all horses except those used for heavy draught. All horses that are required to go faster than a walk are injured by calks. Farm horses are better without shoes, as are all others, but if you MUST SHOE, then use a thin, flat shoe, so that the frog will come in contact with the ground. The shoe should be so made and fitted as to bear upon all parts of the sole and wall that are calculated to bear pressure, and I am of the opinion that the sole as well as the wall is intended to perform its portion of the weight-bearing function; the frog allowed to come in contact with the ground to prevent concussion, thus the weight of the animal is diffused over an extended surface and not limited to the wall alone, as is the common way of shoeing. Slipping is prevented by the rim of the shoe and the wedge-shaped frog grasping the ground. The shape of the frog is such that in the bare-foot horse we have an expansion and contraction of the heel at every step the horse takes. As soon as you place a high heel shoe on the foot, lifting the frog from the ground, you prevent the frog from performing the office which nature designed it, thus causing atrophy, fever, inflammation, contraction, thrush, corns, quarter cracks, etc   In fact 90 per cent. of all unsoundness is the result of shoeing, and no man can shoe a horse except he injures the foot to a greater or less extent. Now, allow me to repeat, in the first place, do not shoe at all; secondly, if you do, let it be a very thin, light shoe just enough to supplant the deficiency of nature; do not load the animal down with iron, straining the tendons as well as injuring the foot.

QUARTER CRACKS.

Are generally found on the inner side of the foot and are most frequent in the forward feet. Sometimes, however, they are found in the middle or other parts of the foot. They may or may not produce lameness, although I have seen some very bad cases of lameness arising from cracked hoofs.

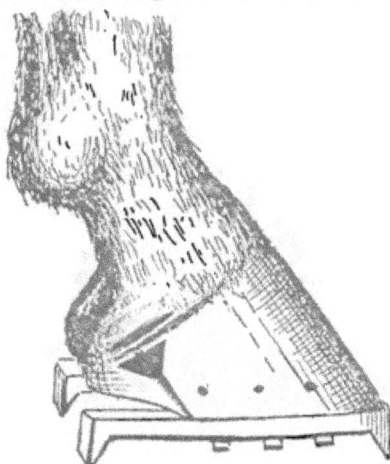

[QUARTER CRACKS.]

TREATMENT.

First clip the hair off around the coronary band for an inch and a half or two inches wide, then you will (put a twist on the nose) take a sparp knife and cut through the skin for a distance of an inch each way from the crack just above the hoof at the junction of the hoof and hair, as indicated by the dark mark on the engraving; be sure to cut entirely through the skin (you need not be afraid of the blood as it will not bleed much); you will now take a flat, red hot iron and burn the edges of the wound enough to sear them over, after which you will remove the twist and take a sharp shoeing knife and pare away the hoof immediately below the cut in a V shape, commencing about one and one-quarter inches from the hair and gradually widening each way from the crack until you have cut one and one-half inch wide at the hair. This

forms a V removing the pressure from the parts, often relieving the lameness at once or in a day or two, if you are thorough and cut the V deep enough, which will be when the
blood starts. If you intend to allow the animal to go without
shoes rasp away the hoof at the heel on the affected side as
shown in the engraving. But if the animal is to be shod, the
shoe should be cut off at the quarter wearing only a *three
quarter shoe*, or you will have to cut away the hoof at the heel,
as shown in the engraving, and apply the common shoe. Be
sure to remove all pressure from the heel and keep it away
until a new hoof is grown. You will now wash the foot to
the fetlock perfectly clean and allow it time to dry, then
apply the fly blister, (See Index) rubbing it in well all around
the coronary band where the hair has been clipped, being
careful not to get any of the blister in the hollow of the heel,
as it will be troublesome to treat after it is once made sore on
account of the quick action of the joint. Allow the blister to
remain for 48 hours undisturbed, then wash it off with warm
soap suds; when dry grease with sweet oil or lard for three or
four days and repeat the blistering, washing and greasing
every six weeks until you have a new hoof, which will be
much thicker and stronger than the old one. If properly
treated this is a *sure cure*. Should the crack be a bad one
and the opening large, you had better pour in a little *hot* tar,
which will remove the soreness. Be sure to keep all pressure
off of the affected heel (as well as all other heels) for pressure
is the most fruitful cause of quarter cracks, especially where
the shoe is allowed to rest on the wall with no sole pressure.
Cracks in any part of the foot are to be treated the same except with regard to shoeing. A toe crack should have the

pressure removed from the two first nails on either side for-forward.

CORNS.

A corn is the result of a bruise, the first appearance is a dark, "blood shot" spot in the triangular space included between the bars and wall at the heel of the front feet and usually the inside heel, although a horse may have a corn on both heels of the same foot. Corns take on a variety of changes; they may change to a horny tumor, or they may suppurate (form matter), or the inflammation may extend to the lateral cartilage (soft part of the heel) terminating in a bony tumor commonly called a sidebone.

Causes.—Shoeing appears to be about the only cause of corns, and in my estimation the common concave seated shoe is the most insane invention that man's brain ever invented. It bears upon no part of the sole except upon the part that is incapable of bearing pressure It is scooped out or made concave all around the foot except an inch or so at the heel, the result of which is corns. A horse with corns is considered unsound although they may not at the time cause lameness.

TREATMENT.

If the foot is hot and feverish remove the shoe, soak the foot in hot water for an hour or two and then poultice.

THE OIL CAKE POULTICE.

Take of the *oil cake meal* one quart, warm water to make a thick mush, a piece of coffee sack about two feet square; spread it out on the floor and place the mush in the center, pat it down a little hollow and pour one ounce of spirits of turpentine in this hollow, then set the foot on top of the mush and gather the coffee sack up around the leg and tie a stout string or buckle a short strap between the fetlock and hoof tight enough to keep it in place. Allow the poultice to remain for 24 hours when it should be renewed. A poultice should have warm water poured over and down the leg into it once every two or three hours, as warmth and moisture are its essential parts.

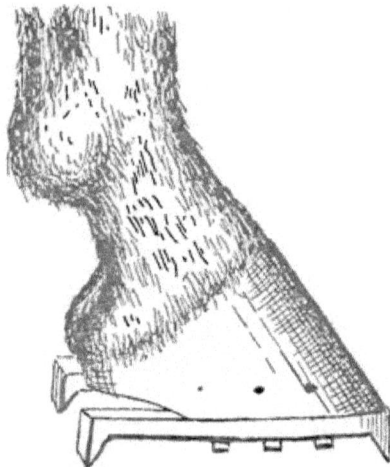

THE PROPER WAY TO SHOE FOR CORNS WHEN YOU ARE
COMPELLED TO USE CALKS.

You will now remove the poultice and pare off the wall and sole alike until you have removed the corn, do not dig a hole into the foot leaving the wall and bar to form a funnel shaped opening to receive and hold all the dirt and filth that comes in contact with the foot. I have seen horses die of lockedjaw from this cause. Cut away the wall from the quarter back to the depth of the corn, as shown in the engraving. If the corn has suppurated a free opening must be made, after which you are to syringe it out with spirits turpentine and then pour in a little hot tar, after which you may put in a small piece of cotton; enough to keep the dirt out, but not enough to fill the opening, causing pressure. You may have to repeat your poultice now for two or three days, or even longer in bad cases. If the hoof becomes hard and dry after you remove your poultice clip the hair and apply the fly blister around the top of the hoof two or three inches wide. Allow it to remain untouched for 24 hours, then wash and grease in the usual manner.

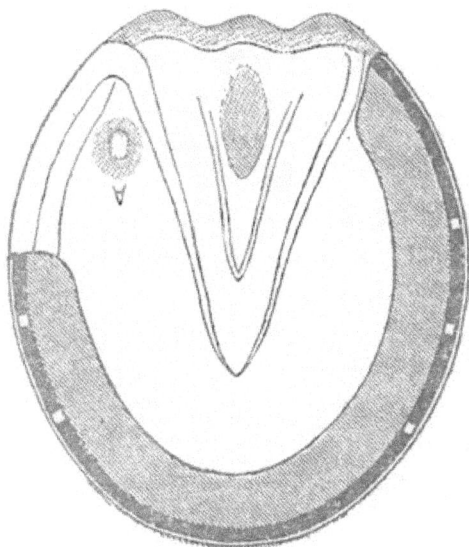

[THE THREE-QUARTER SHOE.]

The radical cure is to be effected by proper shoeing and the three-quarter shoe is certain and speedy, and recommends itself to all unbiased practical men.

INTERNAL TREATMENT.

Give a physic ball at once in all bad cases followed by one tablespoonful of nitrate of potash three times a day in the feed for three or four days.

OSSIFIED LATERAL CARTILAGE.

ENLARGED HEELS, SORE HEELS, HARD HEELS, LAME HEELS.

The cartilage which forms the soft structure of the heels situated above and on either side of the frog frequently take on a bony growth. Only one or both heels may become involved, sometimes a slight enlargement, and at other times it assumes an enormous size. I removed an ossified lateral cartilage for Mr. Horace Merrick, of Big Springs, Mich., from the heel of a large Norman mare, that was nearly as large as the foot itself, with good recovery.

Causes—Are hereditary, predisposing, and shoeing with high healed shoes. It is generally admitted by the best authority that side-bones are hereditary, and that high-heeled shoes are also a cause, for the shock received by the heels when the foot strikes the ground is transmitted directly to the cartilage. Secondly, because the pressure upon the wall at the the heel is unnatural and excessive, the frog being raised from off the ground it forms no support. An injury to the soft part of the heel is liable to terminate in a side-bone, corns have been known to terminate in side-bones, one horse stepping on another's heel is a fruitful cause, barbed wire cuts are very apt to terminate in side-bones, speed horses that over-reach, (jump onto themselves), bruising their heels are liable to side-bones.

SYMPTOMS.

In examining for side-bones you should press upon the cartilage, which is soft, yielding and elastic in the healthy foot, but becomes hard, unyielding and more or less enlarged. Side-bone lameness differs from ring-bone and most other forms of lameness, the toe of the foot first touching the ground·

TREATMENT.

If but one heel is affected apply the three- quarter shoe, if both heels are affected apply a bar shoe, rasping away the hoof from the heel to the first (back) nail as for corns or quarter cracks, freeing the foot from the shoe at least one-fourth of an inch. You will now clip the hair off of the enlargement and apply the mercurial blister (See Iudex) freely, rubbing it in well. Allow the blister to remain untouched for 48 hours, then grease with soft oil or lard. Repeat this blister once in three weeks until you have the desired effect. If there is any lameness remaining after three months you will take the sharp pointed firing iron, first strapping up the well leg, (do not allow any one to hold it, but strap it up), put a twist on the nose or ear, then take the iron at a white heat and burn from three to five holes into the enlargement quite deep—say three-eighths to one-half inch deep —now remove your strap and twist and rub the holes you have burned full of the mercurial blister and repeat your blister as before. This treatment will cure the majority of cases, but rarely we find a case that this treatment will do no good ; then we must resort to *Neurotomy. This is a very

---

* Dividing the nerve which removes the sense of feeling from the affected parts.

simple operation for a qualified veterinary surgeon to perform, and if the foot be a good one will give satisfactory results. I have always been careful in my selection of feet, and have never seen any bad results following Neurotomy. A flat-footed horse should never be nerved, and one that is nerved should be driven carefully over rough, stony, or hard frozen roads, as he is entirely deprived of the sense of feeling in the nerved foot or feet.

NAVICULAR DISEASE.

COFFIN JOINT DISEASE, POINTING THE FEET, ETC.

This is a disease upon which a whole volume might be written, and is, of truth, a source of great annoyance to the practitioner as well as a great pecuniary loss to the horse owner. It usually developes itself slowly, consequently the owner is not much alarmed until it has passed beyond the reach of medical aid. The first stage is that of inflammation, and all efforts should be used to stop the inflammation before an alteration of the structure has taken place. This structural change is indicated by what is commonly called contraction.

CONTRACTION.

Contraction is not a cause, but the result of disease; a wasting away of the structure contained within the horny box. Many contribute contraction to standing on a hard, dry floor, etc, but from actual observation I find that the bare-footed horse, although compelled to stand on the dry hard floor is rarely affected with contraction. Hence I have come to the conclusion that the majority of all contracted feet are the result of shoeing with high calkings. This throws the frog into disuse, in consequence of which it dries up and wastes away the same as any other organ that is thrown into disuse.

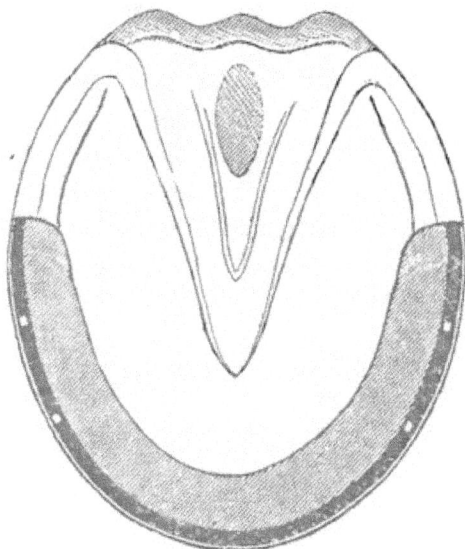

[THE TIP SHOE.]

THE PROPER WAY TO SHOE A FOOT FOR CONTRACTION.

For an example a horse that has a spavin of long standing bearing his weight upon the sound leg with the lame leg flexed, throws the muscles of the hip into disuse, which gradually waste away. This is also true of the shoulder in foot lameness. Contraction does not cause lameness; on the other hand nearly all foot lameness causes contraction.

SYMPTOMS.

The lameness is manifested in two ways: First—Suddenly and perhaps without apparent cause; very often after the horse is newly shod. It is then attributed to some fault in the nailing. After a while the lameness may dissappear and in an indefinite time reappear either in the same foot or in its mate. Second—By a slow process in one or both feet, and this is the most common form, the first noticeable sign being pointing of the foot followed by lameness, which may be of a transient character. For example a horse may go from the stable apparently sound, but the driver occasionally imagines that he is going a trifle lame. The foot is examined but nothing found, he is driven again and again with the same results, until the lameness becomes quite noticeable. If at this period the articulating surface of the joint becomes involved we have

NAVICULAR ARTHRITIS.

A CHRONIC INFLAMMATION OF THE BONY STRUCTURE.

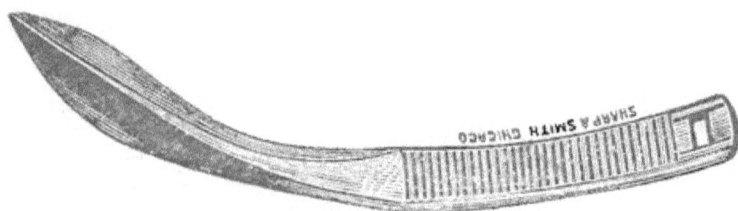

[THE SEATON NEEDLE.]

Remove the shoe and then *poultice* for two or three days then wash the foot and rasp off the bottom until you have it in shape and apply a blister all around the pastern from the hoof up nearly to the fetlock joint being careful not to blister the hollow of the heel, allow the blister to remain from 24 to 36 hours, then wash it off once a day for two or three days and oil the blistered surface with sweet oil or soft grease.

With from six weeks to two months' rest this treatment will cure all mild cases. Should you have a severe case you will add to this treatment the following: After the foot has been trimmed and poulticed as reccommended above, just before applying the blister you will nail a common shoe on (be sure to free the heels), then you will pass a seaton (commonly called a rowell) down through the frog with a curved *Seaton Needle* (See Engraving). Insert the needle at the hollow of the heel, passing down through the frog, coming out about the middle of the frog. Pare away the frog until the blood starts at the place where your needle is to come out before you insert the needle. Pass your needle *down* through the right, and *up* through the left foot. Put a twist on the nose, and if you are a little quick you can pick up the foot and pass the needle through before the horse knows what you are about. Tie the ends of the seaton loose enough so that you can move it a little every day. If the string is too long he will be liable to step on it with the other foot. Allow the frog seaton to remain for three or four weeks. After you have put in the seaton you will apply the blister as already described. After you have washed off your blister and oiled for three days you will again wash with hot soap suds perfectly clean and bathe the blistered surface once a day with "White Lotion" (See Index), using it on the seaton and all about the foot until the seaton is removed.

INTERNAL TREATMENT.

Give a laxative ball or drench at once followed by a bran mash in all cases and in bad cases give a tonic in addition, and the following is a good one: Powdered nitrate of

potash, one pound ; quinine, six to eight drams, rubbed together. Give one tablespoonful three times a day in feed or
on the tongue.

THRUSH.

A discharge from the frog which smells very offensive.
The center of the frog is the part usually affected, and if allowed to run it will spread over the entire foot.

Causes.—The external causes of thrush are standing in
filth and being allowed to run in a wet, muddy paddock or
barn yard. When arising from *external causes* it is easily
cured.

TREATMENT.

Wash with warm water, cleaning out the frog and bars
nicely, then apply a little nitric acid, working it well down
into the diseased parts with cotton batting. Remove the cotton
after a day or two, and should there be any sign of the disease
remaining sprinkle in a little calomel, tamping it in with
cotton as before, observing cleanliness minutely. When arising from *internal causes* such as swelled legs, greese, etc., the
treatment is sometimes tedious.

INTERNAL TREATMENT.

Give a scalded bran mash followed by a laxative ball or
drench. Then take of iodide of potash, four ounces ; crystal
iodine, two drams ; water, one pint. Put the medine into a
bottle, then pour in the water, shake for a moment and it is
ready for use. Of this give one tablespoonful in a feed of
dry oats and bran, well rubbed together, three times a day.

After you have given one prescription wait a week or two, and should you see no signs of improvement repeat this treatment verbatum until you have the desired effect. This treatment, along with the *external treatment* already given, has never failed me in my extensive practice.

CANKER OF THE FOOT.

Heavy horses are more susceptible to this disease than lighter ones. All are, however, alike liable to it from the same cause, viz: Punctured wounds, gravel in the foot and bruises of the sole and frog. It is usually confined to one foot, but if the animal is predisposed to canker, two, three or all feet may be affected.

SYMPTOMS.

A short time after the foot has been wounded in any manner you will notice a fungus growth appear, which protrudes from the wound ; it is vascular and bleeds easily ; the horse will be exceedingly lame. It may extend over the whole bottom of the foot. The sole is absorbed and the foot contracts rapidly, the lameness increasing daily until the horse will only touch the toe to the ground.

TREATMENT.

Remove all the horn or hoof near the canker, then remove all the fungus growth that you can with the knife; do not be afraid of the blood, as it will stop when you apply a dressing of nitric acid. You are to apply a little nitric acid to the sore once a day until it stops growing, then dress with hot tar every day or two. If the lameness is very great you will apply a good blister all around the foot from the hoof half way to the fetlock except the hollow of the heel, and it is a good practice to blister in all cases, as it stimulates the growth of the hoof and hastens the recovery. In all cases you are to remove all of the detached horn, which may include the frog, or it may include the frog and sole both, and should you remove both you are to apply a dressing of warm tar and tow, oakum or cotton, over which you are to tie a piece of stout cloth, or what is better, cut out a piece of tough wood the size and shape of the foot one inch thick and nail on an old boot leg forming a boot or shoe, making holes for strings to lace it up like a shoe. This can be removed easily, which should be done as often as may be deemed necessary, and I do not think it is best to interfere with such a sore too often, as the dressing excites an abnormal growth—just the thing we must here overcome. Any unhealthy or red looking spots are to be touched over with a feather dipped in nitric acid as often as may be necessary to keep them from growing out beyond the surrounding parts.

INTERNAL TREATMENT.

If in good condition give a laxative ball or drench at once, and a scalded bran mash every day or so. Give of the following one tablespoonful two or three times a day : Powdered starch, one pound ; arsenous acid, two and one-half drams, rubbed well together, and it is ready for use. This you will rub up in a feed of oats and bran three times a day. Along with this treatment you must use a good deal of common sense, a great deal of patience, with an undue amount of perseverence and you are pretty sure of a cure. Prof.

Williams recommends removing the entire sole in moderately bad cases, but I do not think it necessary only in very severe cases. It is best to give moderate exercise if the animal is not too lame, and in the worst cases he should have a large, loose box stall, and in no case be compelled to stand tied up in a narrow stall. The cathartic ball should be repeated once every ten to fourteen days.

PUNCTURES—(NAIL IN THE FOOT.)

Punctures from nails may or may not be serious according to their location and depth. Sometimes it only penetrates the sole and does not cause much lameness, and you will have some difficulty in finding what is wrong. Not long since a gentleman brought a lame horse to my infirmary that was lame, as he said, in the *right* front foot, but he could not tell what was wrong. On examination I found a nail in the *left* foot. Another had been fired for a spavin and an examination revealed a small wire nail an inch and a half long in the frog. I could give many such instances, but this will suffice to give you a pointer.

SYMPTOMS.

A horse with a nail in the hind foot will knuckle, and if it be in the front foot will usually point, and when the weight is thrown on the lame foot will bring the other forward quickly.

TREATMENT.

Examine carefully with the knife, take a hammer and tap lightly on the foot, holding it in your hand so that you can feel the slightest move should he evince more pain in one place than another. Cut down carefully until you come to the soft part of the foot, which you will open carefully. Should you find matter there, enlarge your opening, and inject a little spirits of turpentine; this being a thin liquid it will penetrate all parts of the diseased structure and set up a healthy action. After you have washed it out with turpentine you may pour in a little hot pine tar. It sometimes happens that the nail is drawn out when the horse makes the first step after placing his foot upon it. The wound suppurates, pus forms and you find the conditions just described. Should there be much lameness and heat you will place the foot in hot water for two or three hours, then poultice. Repeat this once a day until the lameness subsides, using the *oil cake meal poultice.* (See Index). Make a good sized opening that there may be a free discharge from the wound; occasionally there will be a fungus growth protruding from the wound, and then you have canker. (See Index.)

GRAVEL IN THE FOOT.

It sometimes happens that a small grain of sand or gravel will get lodged between the wall and sole of the foot. This may occur at any point of union of the wall and sole.

SYMPTOMS.

The lameness is quickly developed, and the general symptoms do not differ much from those of a *nail in the foot.*

TREATMENT.

Gravel is to be treated the same as a nail in the foot. If you look carefully you will be likely to find where the gravel entered the foot by following the light colored line around the margin of the foot that marks the union of wall and sole, as it usually leaves a dark spot which you will follow with your knife and probe, and a horse nail makes a very good probe for this purpose. After you have made an opening you are to poultice, following the treatment in the preceding article.

In *paring* and *searching* lame feet great care should be taken not to make them bleed, as the blood obscures the operation and makes it difficult to follow a small spot of discoloration. When the parts are wounded with the knife they are liable to suppurate and cause the healing process to be retarded. In some cases it may be necessary to apply a bar shoe for a few days to remove the pressure from the diseased parts assimilating it over the frog and healthy part of the foot, and leather soles are at times very useful for this purpose.

QUITTOR.

A fistulous wound at the coronary band usually located at the quarters near the heel. Sometimes the heel and quarter becomes wonderfully enlarged, hot and painful, the animal lying around until his body is covered with sores.

Causes—Are corns, pricks from shoeing, gravel, picked-up nails, bruises, etc.

TREATMENT.

If from any external cause make a good opening at the bottom of the foot, wash it out with a syringe and warm water from above and below, then inject one or two tablespoonfuls of the following: Powdered corrosive sublimate, one dram; water, two ounces; hydrochloric acid ten drops; be sure and force this in each and every opening, then apply the oil cake meal poultice. (See Index.) Repeat the poultice daily for three or four days then wash and blister with the fly bl'ster. This will generally effect a cure. Should this fail you will have to enlarge the openings by inserting a sharp knife to their bottom, then introduce a small piece of lunar caustic into each sinus (pipe or opening) as far as possible, and apply a "Wood Ashes Poultice."*

THE WOOD ASHES POULTICE

Is made as follows: Take a piece of coffee sack two feet square, spread it out on the stable floor, then take two quarts of nice, fresh ashes from the stove, wet them with warm water sufficient to make a mush, place this on the coffee sack and put the diseased foot into it; be sure that the ashes covers the sore. This is to remain 24 hours, then remove, wash and apply the oil cake poultice. You must use the ashes poultice carefully as it is liable to blemish by destroying the skin if allowed to remain on too long. This treatment has been very successful. However, quittor is a disease that requires patience and perseverence. Should this treatment fail to produce the desired effect you will be obliged to call in a qualified veterinary surgeon and have him operate by cutting out the whole diseased quarter. This operation requires the use of chloroform.

RING-BONE.

This is a name given to a boney enlargement found upon the lower and upper pastern bones just above the hoof.

---

* Note.—While my own experience with this poultice is somewhat limited, Mr. H. L. King, of Benton Harbor, who has devoted much time to the treatment of lameness, says that this remedy has always been successful in the treatment of quittor.

Ring-bones are of two kinds—true and false. The false ring-bone might be called a splint, as it is of the same nature. When very large it causes lameness, but as a rule it causes no inconvenience, and should not be looked upon as an unsoundness.

TRUE RING-BONE.

Is an unsoundness in every sense of the word, causing great and sometimes incurable lameness. Ring-bones are not the cause, but the result of disease.

Causes.—Allowing the young colt's foot or hoof to grow out long. In the winter season they are either in the soft snow or in a well bedded stall so that the wear of the hoof is comparatively nothing, consequently they attain a considerable length, which throws the weight back on the pastern joint causing an inflammation, exudation, and ossification, which means a ring-bone. This, I think, is the cause of the majority of ring-bones on young animals. However, some are predisposed, inheriting them from their ancestors. An injury of any kind on the pastern joint is liable to terminate in a ring-bone.

SYMPTOMS.

The animal is usually lame for a time before the enlargement makes its appearance, due to an inflammation of the synovial membrane. With a ring-bone on the front pastern, in motion the heel touches the ground first, but when in the hind pastern the toe always touches the ground first. The lameness comes on slowly and will disappear to a great extent on driving.

TREATMENT.

If the animal goes on his toe have him shod with a *spring heel shoe,* not a long calked shoe, but a SPRING HEEL SHOE, which is very thin at the toe and gradually thickened until it is an inch thicker at the heel than at the toe. Should he put the heel down first reverse your shoe and make it very thin at the heel and thick at the toe. After he is shod put him in a loose box stall and give him a laxative ball or drench followed by a scalded bran mash, withhold the hay for 12 hours; give plenty of water to drink. If the bowels are not opened in 24 hours give one-half pint of raw oil. Bathe the affected parts four or five times a day with the white lotion for a week. Should the lameness continue after a week clip the hair off from the ring-bone nearly to the fetlock joint and apply the mercurial blister according to directions. Be careful about blistering the hollow of the heel, as it is troublesome to cure on account of the quick action. Repeat the blister once every three or four weeks until you have made three or four applications. This treatment cures the majority of ring-bones. Should the lameness remain and the enlargement continue to grow you will have to use

THE POINTED FIRING IRON.

Take the horse into a blacksmith shop, put a twist on the nose and have an assistant hold it. If it be a lame front foot strap up the well leg with a foot strap; do not hold it, but strap it up.   If the hind leg is to be fired put on

THE SIDE LINE AND TWIST.

Take a rope 10 or 12 feet long, tie it around the pastern of the well leg, pass it between the front legs over the neck down behind the front leg, wind it once around itself, draw the hind leg forward far enough to compel him to stand with his whole heft on the lame leg, and have an assistant hold the rope. You will now proceed to fire him. Have your iron as hot as you can get it without melting, make about three rows of dots around over the surface of the ring-bone about one inch apart; if you burn them close together they will be likely to cause a blemish. When the enlargement is great burn deep enough to penetrate the bone. Burn from eight to twelve holes; do not burn too near the hoof. Now remove the fastenings, put him into a nice clean stall, tie his head so so that he cannot get to foot and apply the *mercurial blister* over the entire ring-bone, rubbing the holes full of the blister.

Allow it to remain undisturbed for 48 hours when it is to be washed off and cared for in the usual manner. If there is much pain give a dose of physic followed by a scalded bran mash, etc. In treating a ring-bone on the hind leg you must always apply the *spring heel shoe* on the well leg to remove the extra strain caused by favoring the affected leg. A horse suffering from any kind of lameness in the hind leg should not be compelled to stand in one position for any great length of time without a sling under him, as this is itself a fruitful cause of coronitis* and laminitis.† Allow him a loose box stall in preference to turning out to pasture for the first four weeks of treatment, then allow him a six or eight weeks' run on grass. Repeat the blister once in four weeks until you have the desired effect.

BONE SPAVIN.

A spavin is an enlargement on the inner and lower part of the hock joint, and is the result of an inflammation of the gliding surface of the bones resulting in the complete union of two or more bones. Spavins are the cause of great, and frequently incurable lameness. The lamness of bone spavins is as a rule curable in the young and middle-aged,

[A BONE SPAVIN.] and incurable in the old horse.

Cause.—Anything that will set up an irritation of the articulating surface will cause a spavin. Prof. W. Williams, in his work on surgery, says that "shoeing with long calkings is one of the most fruitful causes of spavins," and I quite agree with him, and will add that the way some plates are

----

* An inflammation of the corona band, which is the union of the hoof and hair.

† Founder.

applied causes spavins. The outside wall of the hoof is
thicker and stronger than the inside wall, and the average
shoer makes the shoe to narrow at the quarter, nails it on the
foot and rasps off the wall to fit the shoe; this weakens both
quarters, but on account of the difference in their thickness
the rasping has weakened the inside so that there is no bearing at all on the quarter, the shoe resting on the toe and heel.
The heel soon wears away, and should it withstand the excessive weight and wear until the next shoeing it is cut away to
get a bearing for the new shoe. By this time the foot is
rolled in sufficiently to cause great strain upon the inside of
the hock joint, which may result in a spavin. An injury to
the foot sufficient to cause the animal to stand on the well leg
for 10 or 15 days is very liable to cause a spavin. For this
reason it is always best to use a sling in such cases. An injury to the hock as a kick or blow may terminate in a spavin.

SYMPTOMS.

Spavin lamenes, at first is of a transitory character, will
show itself when the animal first comes from the stable, but
soon disappears. The lameness usually precedes the enlargement, although we sometimes have quite an enlargement with
little or no lameness, and should a horse having a large spavin
be taken suddenly very lame you had better examine the foot
closely for a nail or gravel before venturing an opinion, as I have
seen horses fired and blistered for spavin lameness when the cause
was in the foot. The horse having a spavin will evince more
lameness when compelled to stand around in his stall one way
than when standing around the other. Make him stand
around several times each way watching him closely. If you
are not satisfied take him out and walk him a little, noting
carefully the amount of lameness shown, stop him, take up his

lame leg in about the same manner that a smith would to nail on a shoe, making the bend at the hock as short as possible, hold the leg in this position for two or three minutes and have hime led off as soon as you let go of the leg. If there is a spavin he will show a great deal more lameness for a few steps than before the test. A spavined horse drags or scuffs the toe of the lame foot while in motion, and usually rests the leg on the toe while standing. In cases of long standing the hip is usually atrophied (shrunken) away. Stand about two feet from the horse's head, first one side and then the other, looking at the hocks, comparing them carefully, and you will be likely to detect any irregularity. A horse may have a spavin without any enlargement.

TREATMENT.

It is impossible to tell whether you can cure a spavin or not' You cannot remove the spavin—so far it is incurable—but if we can get them to go sound it is what we call cured. First put the foot in shape by removing the shoe and paring it so that the hoof at the heels on each side are alike. Put a spring heel shoe on the well foot and turn him into a loose box stall. Give a physic and scalded bran mash, withhold the hay for 12 hours, bathe the hock with white lotion for two or three days, then clip off the hair and apply the *mercurial blister* if it be a mild case without much enlargement, but if there be considerable enlargement then fire at once.

HOW TO FIRE A SPAVIN.

Take the animal to a blacksmith shop put the twist on and attach the *side line* Page 100, to the well leg. Now take the pointed firing iron Page 99, at a white heat, and burn from six to ten holes. Keep the iron very sharp and burn deep; penetrate the bone for a quarter of an inch. Return him to his box stall and apply the mercurial blister, rubbing the holes full. Allow the blister to remain undisturbed for 48 hours then wash it off with warm soap suds once a day for two days, then grease thoroughly with sweet oil or lard once a day for three days. Repeat the blister once in three or four weeks until you have the desired effect. *Remember* that those who have become successful in the treatment of *lameness* have done so by devising means of keeping the animal *quiet*.

TREATMENT OF PARTICULAR SPAVINS.

You will sometimes find a large spavin with a groove running across it just in front of the "wart" on the leg. This groove is caused by a tension of the cunean tendon, which passes over the seat of the spavin and is attached to the cuneiform bone, as shown by the accompanying cut. When you find such a spavin, secure the animal in the usual way. Take the *feathered firing iron* (See Index) and burn one stripe about an inch back of the large blood vessel that runs down the inside of the leg. Make the stripe one and one-half

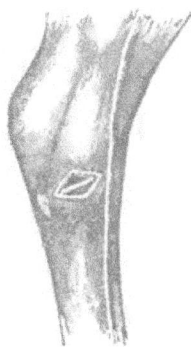

[CUNEAN TENDON.]

inches long running up and down the leg across the groove, burning deep enough to cut this tendon in two. As you are burning the skin will draw back so that you will see the tendon before it is divided; it is white in color and three-eighths of an inch wide. After you have the tendon severed you may burn a few deep holes with the pointed iron, as already described, and treat as in other spavins. This treatment has often proved successful when all others have failed.

CURB.

This is a sprain to the calcaneo-cuboid ligament causing an enlargement at the back part of the hock joint about four inches below the point of the oscalcis (where the "hamstring" is attached to the hock). At times they attain considerable size, again they may be very small, and the amount of lameness may not appear to depend upon their size, as I have often seen large curbs that seemed to give little or no inconvenience; at other times a very small curb would cause the animal to go "dead lame." Curbs of recent standing usually cause lameness especially in the young animal. A horse lame from a curb stands with the leg flexed or extended under the body. Two or three days' rest often removes the lameness, but it will return when put to work again. Curb lameness increases with exercise.

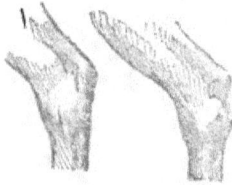

TREATMENT.

Have the animal shod with the *spring heel shoe*; this removes the strain from the tendon. If hot and feverish bathe with the *White Lotion* for a few days, then clip the hair and apply the *mercurial blister*. If any enlargement remains after you have made three applications in the usual manner of blistering you will have to use the actual cautery.

HOW TO FIRE A CURB.

Secure the animal in the usual way with the *side line*, then take the feather edged iron and burn one stripe down

the center of the leg the length of the curb, then burn one
line on each side of this one about two-thirds as long and
three-qarters of an inch from it running parallel with each
other.   Burn quite deep, but not deep enough to cut through

[THE FEATHERED FIRING IRON.]

the skin.   After three days apply the *mercurial blister* ; allow
it to remain two days, then wash once a day for two days with
warm water and soap, then grease with sweet oil for three
days, then wash again with soap and warm water and bathe
two or three times a day with the *White Lotion* for a week or
two.   This treatment is usually successful.

THE FLY BLISTER.

Powdered Spanish Flies.................... 4 ounces
Rosin. Spts. of Turpentine and Beeswax, each 2   "
Lard.................................16 ounces

Mix.   Melt altogether over a slow fire and stir until cold
when it is ready for use.   Clip the hair off and apply once,
rub it in for 10 minutes, allow it to remain for 48 hours then
wash with warm soap suds.   When dry grease with sweet oil,
lard or other soft grease.

THE MERCURIAL BLISTER.

Fly Blister............................... 1 ounce
Red Iodide of Mercury.................... 2 drams

This is an excellent blister for the treatment of bony and
fibrous growths, callous lumps, wind galls, puffs, ring-bones,

spavins and enlargements or chronic swellings on any part of the horse and is to be used the same as the fly blister.

HOW TO BLISTER.

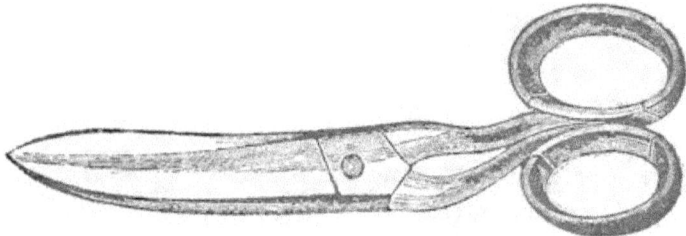

Clip the hair off of the entire surface to be blistered, then wash it perfectly clean, allow it time to dry and apply the blister pretty freely once. Rub it in for 10 minutes allow it to remain untouched for 48 hours, then wash with warm water and soap. When dry grease it well with sweet oil, melted lard or some other soft grease once a day for two or three days.

To obtain the full effect of a blister a quantity of the ointment should be laid on after the rubbing in is completed.

GENERAL REMARKS ON BLISTERING.

Never blister more than two legs at one time and three weeks are to elapse before the others are blistered.

HOW OFTEN.

No blister should be repeated oftener than once in three or four weeks, and do not blister too large a surface at one time, as you are apt to get up a constitutional disturbance. In this case you must wash the blister off at once perfectly

clean, oil the surface with sweet oil and give in a half cup of
water a tablespoonful of bi-carbonate of soda three or four
times a day.

SWELLING.

Sometimes blisters, no matter how carefully they are applied, will cause extensive swelling with a tendency to suppurate. This is due to the animal's general bad health bordering
on erysipelatous diseases. Treatment consists of purgatives
and diuretics* internally. Wash the affected parts perfectly
clean with warm water and soap, then bathe freely every two
or three hours with the *White Lotion.*

TO PREVENT ACCIDENTS.

Fix the horse so that he cannot bite the blistered surface
or rub it with his lip and nose, and thus blister the mouth
also. Some horses are nervous and will rub themselves. This
must be prevented as it is apt to cause a blemish.

DREAD OF BLISTERING.

The average man has a perfect dread of blistering. You
just say to a man, "That horse needs blistering," and the
chances are that he will say "I'll wait awhile and see if he
will get better without it, for I do not want him blemished."
How absurd to think a blister cannot be so managed as to
leave no mark. Whenever your horse needs blistering do not
hesitate, but blister him at once and take care of him. Give

---

* That which increases the flow of urine.

him a little needed rest, that which is due any faithful servant, and you will be amply repaid for your trouble.

THE RESULT OF BLISTERING.

"What good does the blister do?" is a common expression. It excites within the already diseased structure a reparative inflammation, or an absorbent inflammation, which hastens the formation of reparative material, by which ruptures are united, ulcers healed and disease removed. Lameness is often removed by blistering, which assists nature in the process of repair, hence it will be seen that it ts necessary to apply blisters in all cases when organic changes in the parts involved are even suspected.

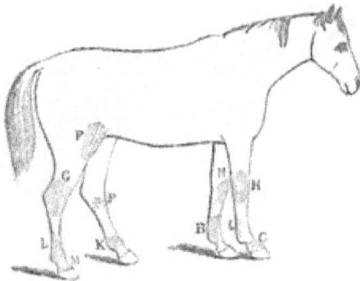

[THE HORSE FIRED.]

Showing the portions commonly fired. F, Stifle; G. Hock; H H, Knee; P. Spavin; M C, Ring-bone; B L, Tendons and Fetlock; J, Back Tendon; K, Fetlock.

ACTUAL CAUTERY

Is much more feared than a blister, and often removes pain and lameness where repeated blistering has failed. Prof. W. Williams, author of "Prin. and Prac. of Veterinary Surgery," says that in bone disease, and in all cases of chronic lameness, it is of great benefit, and acts by powerfully exciting the healing process in the parts diseased. Thus it will be seen that lameness of a bone spavin is removed through the inflammation excited by the firing iron in the diseased bone, increasing a supply of material for the purpose of uniting them together into one immovable mass. For the treatment of thickened and enlarged tendons the actual cautery will give the best satisfaction.

THE STIFLE JOINT.

This joint is not free from diseases, some of which are of a very grave nature. One of the most common affections is a

DISTENSION OF THE CAPSULAR LIGAMENT.

This is recognized by a soft tumor or enlargement in front of and a little below the stifle joint resembling a puff, and is a cousin to the bog spavin family. This enlargement usually but not always causes lameness. However, I am inclined to think that when suddenly developed in horses of mature years, that it always causes lameness.

TREATMENT.

If of recent origin bathe freely four or five times a day with the *White Lotion* (See Index) for a week or ten days, then fire with the *feathering iron* using the design shown on Page 106, marked F. Do not make the lines too close together,—say three-quarters of an inch apart—and quite deep, but not through the skin. You will have to apply the *side line* (See Index) to the well leg. The third day after the firing apply the *mercurial blister* as directed (See Index.) It may be necessary to repeat the blister two or three times to effect a cure. This is a disease that requires rest while undergoing treatment, with a cooling laxative diet.

LUXATION OF THE PATELLA.

DISLOCATION OF THE STIFLE JOINT.

Occasionally, but not often, we have complete dislocation of the patella (the knee cap of the horse situated on the stifle joint—See Skeleton) and when once you have *dislocation* your only chance of a recovery depends upon prompt treatment. You will hear a cracking or snapping sound at every step the animal takes.

TREATMENT.

First have a shoe make to fit the foot with a spur welded on at the toe about eight inches long, turned up a little at the end in sleigh runner fashion, punch a hole in the end of the spur to receive a cord, nail the shoe to the foot, tied a cord to the spur bring the cord up around the neck, draw the foot forward in a natural position and make the cord fast. This shoe is very serviceable in the treatment of many ails of the foot and leg. You will now apply the *fly blister* as directed on Page 106. Allow perfect rest for six weeks or longer.

CRAMPS.

RHEUMATISM OF THE THIGH.

Young animals are often troubled with rheumatic cramps of the thigh near the stifle joint caused by improper food and a lack of exercise. Idleness is the horse's worst enemy, and it is acknowledged that a horse cannot be kept in condition without exercise. This disease has often been dubbed

"STIFLE OUT OF JOINT."

This is not a good name on account of expressing a condition which is not present.

SYMPTOMS.

Perhaps the first thing noticed will be an inability to stand over in the stall. Then you try to back him out of his stall, which seems an utter impossibility. This is a common expression, "he acts as though his foot was nailed to the floor," and if his stall is large enough he will walk around, the diseased leg twisting the foot without lifting it from the floor. It is almost impossible for him to back, and when moving it is with great difficulty that he brings the leg forward, and will not move unless compelled to do so. Excitement will usually cause the animal to throw his weight upon the leg in such a way as to remove the cramp when he will walk off all right. Give him a good, smart cut with a whip flourishing it in a manner to create some excitement. This will remove the cramps from nine-tenths of all the cases you meet. But I have been called to treat a few cases where the cramps remained ridged and unyielding for several days.

TREATMENT.

First give a *physic ball* followed by a scalded bran mash, withholding the hay for 12 hours. If the bowels do not open and the cramps remain after 24 hours give the *turpentine drench* followed by tablespoonful doses of salicylic acid three times a day for three days.

EXTERNAL TREATMENT.

Bathe the thigh once a day with the *hartshorn liniment* (See Index) until it produces a mild blister, then omit. Give plenty of exercise, and remember that exercise ceases when fatigue begins.

THE SPLINT.

This is a common cause of lameness and is usual'y met with in young animals. It is caused by a strain, sprain or bruise, which loosens the periosteum (skin of the bone) forming a sack or cavity which fills with exudate and is at first soft, but gradually hardens until it is bone itself. If suddenly developed they cause lameness regardless of their location, but if developed slowly they cause no lameness unless they are close to and interfere with the action of a joint.

TREATMENT.

If suddenly developed, or if large, the skin of the bone is to be cut open as follows: Make a small hole in the skin at the lower edge of the splint by cutting crosswise of the leg, then take a small, narrow bladed knife and introduce it into the hole made in the skin flatways, push it up between the skin and bone until the point comes to the upper border of the splint, then turn the cutting edge to the bone and withdraw it, cutting down on the bone as you bring the knife out without enlarging the hole in the skin. This relieves the tension or strain on the skin of the bone.

Now apply the *mercurial blister*, Page 106, as directed. Repeat the blister once a month until you have the desired effect. Small or chronic splints are to be treated with the blister alone. Splints usually flatten down with age and do no harm. Splints are not considered an unsoundness unless they are close to a joint or cause lameness.

ATROPHY (SWEENY).

This is a common ail and is usually seen in young horses. Ploughing seems to be the most fruitful cause of shoulder sweeny and spavins or other lameness; the cause of hip sweeny. Occasionally you will see the muscles of the arm or thigh wasted away, which is due to some injury cutting off the supply of nutriment from the parts. In the first stage there is swelling, but it is not often noticed. The first thing usually observed is shrinking away of the muscles. May or may not be lame, but generally is a little stiff.

In the very worst cases a sure cure can be affected although it may require a long time before the muscles will attain their usual size. A loose box stall is chosen in preference to the pasture. Keep the animal quiet for a month or six weeks adopting the following treatment: Foment the shrunken parts with hot water for a while, then rub dry and apply the *fly blister*, Page 106. After the blister has worked and been washed off in the usual manner, bathe the affected parts with the following :

```
Tincture of Camphor......................  4 ounces
   ..        .. Opium.......................  4    "
   ..        .. Arnica......................  4    "
```

Mix and use once a day. If necessary repeat the blister once a month, covering the entire shrunken parts until you have the desired effect.

VETERINARY SURGERY.

CHAPTER III.

Congestion and Inflammation Defined.

By the term congestion we understand a condition caused by, and dependent upon, an abnormal afflux of blood to a part.

Congestions are of two kinds, active and passive, differing from each other in several marked particulars, more especially in relation to their causation and methods of treatment—not so much on the actual anatomical differences between the two. We will endeavor to define the differences existing between active and passive congestions in as few words and as simple terms as possible, remembering that in a great many instances the two may exist in the same organ or part of the body at the same time, or during the course of the same disease, at different times, in varying degrees of intensity.

An active congestion is one which is produced by either the direct or indirect irritation of a part or organ. A passive congestion is one which is not dependent upon an irritation, but upon some mechanical obstruction to the return of the normal quantity of blood from an organ or part of the body.

As simple illustrations of the above let us take, first, a superficial burn. The irritation produced by the heat is fol-

lowed by an increased afflux of blood to the part, and an
active congestion of the part follows, as shown by increased
redness and swelling. Second, let us suppose a string tied,
not too tightly, around the finger. The flow of blood from the
part being thus mechanically obstructed, a passive congestion
results, also acc mpanied by increased redness and swelling.
An active congestion may also be produced by forcing an
abnormal amount of blood to a part different from the part
primarily affected, as, for instance, in the case of long exposure of the surface of the body to cold. Now cold, as every
one knows, when applied to the exterior of the body, contracts
the blood vessels on the surface, and thus drives the entire volume of blood to the interior portions and organs. This, of
itself constitutes a congestion. and those organs will be
affected in the greatest degree which possess the least power of
resistance.

Passive congestion is also frequently produced in the following manner: The heart's action being weakened from fever
or other acute illness, and the natural tonicity of the bloodvessels being loosened by the accompanying nervous depres-
sion, the "vis a tergo" or force from the starting point, i. e.,
the heart, is not sufficient to overcome the force of gravity in
the more dependent portions of the body, and as a result
stagnation of the circulation or passive congestion follows.

These two forms of blood afflux and stasis constitute the
primary phenomena of congestion. At a longer or shorter
period of time the secondary symptoms begin to show themselves. From the foregoing remarks it is plain to be seen
that the only rational way to treat congestion in its primary
stages is to equalize the circulation, relieve the congested
parts or organs from the excessive amount of blood contained
in their structure, and maintain it in this normal condition

till nature has an opportunity to recuperate the over-distended parts. In simple congestion there is no diseased condition of the tissues themselves present, except in few instances, and in these instances the diseased condition must have preceded the congestion.

WOUNDS.

These are of various kinds and are called *incised, punctured, lacerated, contused,* and *poisoned.* The *incised* wound is where the parts are smoothly divided with a sharp cutting instrument, and its length exceeds its depth. Usually, if it is lengthwise of the muscle, it is not very serious, but if crosswise, a great gapping wound and considerable bleeding is the result.

TREATMENT.

First if on the leg and the bleeding is excessive put a good stout piece of cloth or handkerchief around the leg loosely above the cut, then pass a stick through the cloth next to the leg and give it a few turns (as a binding pole on a wagon). Tighten this until the blood stops flowing. You will now bathe with *hot* or *cold* water removing all foreign substances such as dirt and hair. Now loosen the twist until the blood starts; this is to show where the severed blood-vessels or arteries are, which you will draw out and tie with a linen thread. Should you fail in this, lose no time in sending for a qualified veterinary surgeon, as the cord around the leg will soon cause extensive swelling and perhaps gangrene. After the bleeding is stopped bathe the wound thoroughly with the *aqua corrosive lotion* (See Index). The next step,

according to all authors of today, would be to close the
opening by

SEWING UP THE WOUND.

As I have had no little experience in this particular
branch of surgery, I cannot too strongly condemn the use of
of the needle. In the first place the stitches will tear out in
three or four days; if not the chances are that you will be
obliged to cut them out allowing the scrum or pus to escape.
Again when pus is imprisoned it will burrow toward the most
dependant part causing grave results, often forming sinuses
(pipes), which are difficult to treat. Again we often have
blood or matter poisoning follow the closure of wounds by the
needle, that would never have occurred if the wound had been
left open. Not long since I had an occasion to assist a professor of a veterinary college in sewing up a lacerated wound.
While the work was in direct opposition to my understanding
of surgery I said nothing until the professor and myself drove
off, when I ventured to remark, "Professor, why did you sew
up that wound?" "For five dollars," was his witty retort.
Then he continued, "If I had not sewed up that wound the
owner would have got someone else to do it; the stitches will
sluff out in a few days, when he will be satisfied to have it
treated as an open wound." While this may be true to a certain
extent, it appears to me that this treatment is not only useless,
but *cruelty* to our dumb animals, causing not only much pain
and suffering, but death is often the result of such quackery.
There may possibly be a case where a few stitches will be of
service, but the chances are 99 to 100 against you. Having
thus pretty thoroughly condemned what I consider a barbarous practice, we will, as the Irishman says "be getting after
our subject."

Instead of sewing up the wound you will now return the animal to a comfortable stall and give him a scalded bran mash, feeding on a cooling laxative diet. Bathe the wound once a day with *aqua corrosive lotion*. Keep the adjoining parts clean. Be very careful that you do not irritate the wound while dressing, causing it to bleed, as the irritation is apt to start the red unhealthy growth called

PROUD FLESH.

Proud flesh is a red, fungus looking growth and is liable to spring up in a wound at any time during the healing process, and it is to be treated with astringents and caustics; *powdered borasic acid* is one of the best mild remedies that I ever used. *Quinine* is another good remedy. These are to be sprinkled over the wound quite freely. When a stronger remedy is required take *nitric acid* and aply it with a feather. This remedy must be used with much care; just touch the red, firey looking spots, being careful not to get it on the white, healthy looking surface, for when applied to healthy flesh it is likely to set up the action we are trying to subdue. When a wound is healing there will be a light colored line around its outer edge. This line should never be irritated in any way. After the nitric acid has been touched to the sore for 10 or 15 minutes you are to apply a little sweet oil with a feather. Use the acid once a day until you have the desired effect.

PUNCTURED WOUNDS

Are made either by a sharp or blunt pointed object, and and the depth is greater than the length. These wounds are usually to be looked upon as of a very grave character. The flesh being pierced to a considerable depth the pus which forms cannot escape freely. This imprisoned pus gives rise to febrile disturbance, blood and matter poisoning, fistulous openings, etc.

TREATMENT.

The first thing to be done is to enlarge the opening suffi cient to admit your finger or hand to its bottom with which you are to remove all foreign substances, wood, dirt, hair and bruised flesh, then wash out the wound and dress with the *aqua corrosive lotion* or *white lotion.* Should the wound be a deep one you will now take a lump of cotton batting. tie a string around it and introduce it into the wound to its b ttom changing it once a day. The batting absorbs the pus and at the same time keeps the wound open until it heals from the bottom. *Internally.*—You are to give the *turpentine drench* at once, repeating the oil as often as may be necessary to keep the bowels *loose.* Should there be much constitutional dis- turbance and fever, give a few doses (15 drops) of ac nite.

A CONTUSED WOUND

Is a bruise without the skin being cut or torn. This is often seen in the man—"a black eye." There may be a rup- ture of a blood vessel; ec hy mosis is the result. This may run to suppuration, or it may be taken up by absorption. *Speedy cuts, collar boils* and *bruised knees* are included in this class of wounds. They sometimes assume a considerable size and are soft and fluctuating.

COLLAR BOIL.

The contents are to be removed by cutting an opening two or three inches long, make your cut with the hair, and it is usually a half inch through the flesh to the serum, which is a red, watery looking fluid. After you have made the opening introduce your fingers and remove all the bruised tissue that you can, then wash it out with

Zinc Sulphate	8 drams
Sugar of Lead	4 "
Carbolic Acid	1 "
Water	1 quart

Mix and wash the wound with a syringe once a day; you will also clip the hair off of the enlargement and apply the *fly blister* (See Index). You must not let the external opening close until the internal parts have healed or grown up. It can be kept open by inserting two fingers once or twice a day into the opening. Collar boils are sometimes very troublesome to treat. The *aqua corrosive lotion* is also a good remedy, and much benefit will be derived from the blister, which should never be omitted.

BRUISED KNEES AND SPEEDY CUTS.

These are to be treated alike. Give four ounces of sulphate of magnesia once a day for a week or until the bowels are quite loose. Bandage the affected leg and keep it wet for 48 hours with the following:

Saltpetre	4 ounces
Sal Ammoniac	4 "
Common Salt	2 pounds
Water	2 gallons

Mix and keep the leg wet for 48 hours, then remove the bandages, rub dry and apply the *mercurial blister* (See Index) and repeat this once in three weeks until the enlargement is removed. Should the swelling remain soft with a tendency to

point you must open it up with a free incision and dress with
the *white lotion, aqua corrosive* or carbolized water. Repeat
the blister once every three or four weeks until you have the
desired effect.

POISONED WOUNDS.

SNAKE BITE.

Whenever an animal is bitten by a snake, as it sometimes
happens, the swelling will be great and appear suddenly.
You cannot always tell to a certainty that the swelling is
caused by a snake bite, but if the animal has been running
where there are poisonous snakes and the swelling is suddenly
developed in the absence of other symptoms, I should treat for
snake bite.

TREATMENT.

First give the *turpentine drench*, then take ot

Permanganate of Potash.................. 20 grains
Water................................... 2 ounces

Mix and inject into the swollen parts with the *Hypodermic
Syringe*, Page 42. Introduce the syringe needle into the
swelling an inch or two injecting the contents of the syringe
in several places, Then bathe freely with aqua ammonia and
spirits of turpentine, equal parts. This treatment is magic,
often effecting a complete cure in 24 hours.

BARBED WIRE CUTS.

In this particular branch of surgery I have had much
experience with good results, and the following is the mode of

procedure: Secure the animal with the *side line* or *foot strap* as the case may necessitate, apply the *twist* to the nose, then take the *dressing shears* and snip off all the loose threads of tendenous or other substance, take up and tie any bloodvessels or arteries that may seem necessary, then bathe with *aqua corrosive lotion*. If there are any flaps or loose skin cut them off also. After bathing with the lotion you are to sprinkle powdered iodiform all over the wound. Now leave it untouched for 48 hours then bathe with the *corrosive lotion* every other day and the *white lotion* every other day alternately. Do not sew up the wound nor allow anyone to do it for you, as this will pen up the bruised tissues, when it will be liable to terminate in blood poisoning matter poisoning or gangrene. Leave the wound open, and if there are any pockets to hold pus, cut them open to their bottom.

INTERNAL TREATMENT.

Give a scalded bran mash once a day containing four ounces of epsom salts, and feed on a laxative diet. If the wound be a bad one you had best give the following:

Sulphate of Quinine....................... 3 drams
Nitrate of Potash........................... 8 ounces

Mix. Give one tablespoonful three times a day; allow plenty of water. Wounds that look frightful will heal up in a short time without a scar if this mode of treatment is followed. Powdered borasic acid is a splendid dressing for this class of wounds, and is nicely applied with the insect powder bellows.

SCABS FORMING ON WOUNDS.

This is the proper way for wounds to heal, and when a scab forms do not disturb it on an opened wound (one that its length is greater than its depth). Keep the surrounding parts clean, but leave the scabs alone; they protect the sore from the air, and nature will throw them off when she gets done with them. It might be well to spray or sprinkle the wound with one of the lotions.

POCKETS IN WOUNDS.

Sometimes the fresh wound has a pocket; again the pus or matter burrows, forming pockets. In all such cases you are either to make an independent opening from top to bottom or make a dependent one at the bottom, through which you had better pass a stout piece of tape, forming a seaton (sometimes called a rowell). This keeps the hole open so that the pus can escape. A free discharge of pus or matter is essential to a speedy recovery in all wounds.

OPEN PAROTID DUCT.

SALIVA FLOWING FROM A WOUND.

The steno duct winds round the lower jaw bone in company with the artery and vein, and enters the mouth between the second and third upper molar teeth. Where it crosses the jaw it is only covered by the skin, hence it is liable to be opened by a kick or blow, although I never saw one opened except by a bungling horse doctor. But from whatever cause it may be opened the saliva runs from the wound instead of flowing into the mouth. When the animal is not eating the discharge is very slight, but while eating, especially if the food is very dry, the discharge is abundant, for this gland secretes in direct ratio with the dryness of the food.

TREATMENT.

In recent cases close up the wound, but if of long standing you will have to re-establish the opening into the mouth.

This is an important part of the treatment and is done by introducing a seaton from the opening of the duct into the mouth and allowing it to remain from four to seven days when it is to be removed and the external wound closed up with a small clamp, pass two small horse nails through the lips of the wound pinning them together and apply the clamp above them. Tighten it a little each day until it sloughs off which will take from seven to fourteen days. After the clamp has been adjusted you are to give a little dry food to excite the secretion of the gland in order to keep the opening into the mouth patent. Great care must be exercised to keep the external wound closed until it heals up. If the treatment by seaton should prove unsuccessful then the gland will have to be destroyed by injecting the following:

Nitrate of Silver	½ dram
Nitric Acid	1 "
Water	1 ounce

A strong syringe will be necessary to force this into the various ramifications of the gland. This arrests the discharge by exciting an adhesive inflammation of the gland destroying its functions forming a solid indurated mass, which is removed by absorption. In no case are caustics or actual cautery to be applied. In recent cases close the wound with stitches and colodion, forcing the animal to abstain from eating any dry food for three or four days, giving gruel and sloppy food. A good smart blister will hasten the closure if it seems to be healing too slowly. This is a piece of surgery that should be intrusted to a qualified surgeon only.

THE BOG SPAVIN.

A bog spavin is a puffy tumor situated on the inside of the hock joint near or a little above and in front of the seat of the bone spavin, and has always been considered more serious to treat than a bone spavin. However, I have invented a combined *bog spavin and thorough pin truss*, which has proved very effectual in the treatment of this heretofor unsurmountable ail. This truss is so constructed that any novice can apply it, and it will retain its position without injury or inconvenience. The animal can lie down in the stable or be turned out to pasture while wearing it. This truss has a medicated pad that covers the bog spavin, and is so arranged that it will not chafe the leg. It is adjustable in size, will fit a yearling colt or a horse of mature years. Full particulars sent on application. Price, $6.00.

LITHOTOMY.

DIFFICULT URINE, BLOODY URINE, ETC.

Lithotomy is an operation performed for the removal of gravel or other substances from the bladder and is usually followed by favorable results if properly performed. The presence of foreign substances in the bladder is indicated by ineffectual attempts to urinate and when the urine does come it may be mixed with blood. Each attack is more aggravated, and if the cause is not removed death follows, which is usually attributed to colic.

RUMENOTOMY.

Occasionally we find a case of chronic impaction of the rumen which resists all medicinal treatment, and then we have to perform *rumenotomy*, which means that we are obliged to cut open the rumen (stomach) and remove its contents with our hand and we are liable to find a hair ball made from the hair taken while licking each other; we also find nails, pieces of wire, large pieces of cloth; in fact we are liable to find almost anything that has disappeared from the premises in the bovine stomach. Mr. Chauncey King, of Benton Harbor, Mich, at one time found a pocketbook containing $17 in the stomach of a veal calf. When it becomes necessary to open the stomach you will first clip, shear or shave the hair closely from the surface to be operated upon, then tie the cow with her right side to the wall, with three men to keep her quiet. Now plunge a sharp pointed knife (a small butcher knife will do) down into the stomach midway between the last rib and and the point of the hip about four inches from the short ribs that cross the back—not four inches from the center of the back, but four inches from the ends of the short ribs—cut downwards until the hole is large enough to allow a free passage of the hand, then take a spaying needle (See Engraving) and sew the walls of the stomach to the outside flesh and skin; this is to prevent anything from falling into the abdominal cavity while cleaning the stomach. You will now remove all the contents of the stomach with your hand after which you are to oil the stomach with a quart of raw oil, wash the wound perfectly clean with warm water, cut out the stitches and sew up the stomach, first using carbolized cat gut; sew over and over taking the stitches from three-eighths to one-half inch apart, push the ends of the strings down into the stomach when done. Now sew up the outside opening, flesh and skin, with a strong, waxed linen string, take the stitches deep so that they will not tear out, and dress with *white lotion* (See Index). Now prepare the following:

Fluid Extract of Nuxvomica.......... .... 2 ounces
Water................................. ½ pint

Mix. Give one tablespoonful three times a day in feed or on the tongue. This operation is usually attended with favorable results when properly performed.

TRACHEAOTOMY.

This operation is often performed for the immediate relief of horses suffering from acute diseases of the throat such as influenza, strangles and puprura hæmorrhagica. I have also operated upon broken wind horses, "whistlers," with good results. The *modus operandi*: Put a *twist* on the nose and have an assistant elevate the head, now make a cut through the skin to the trachea (windpipe) two and one-half inches long, about midway of the under part of the neck at a point where you can feel the windpipe with nothing but the skin covering it, next cut a round hole in the cartilage by taking out a part of one of the rings, and introduce the *trachae tube*. If you

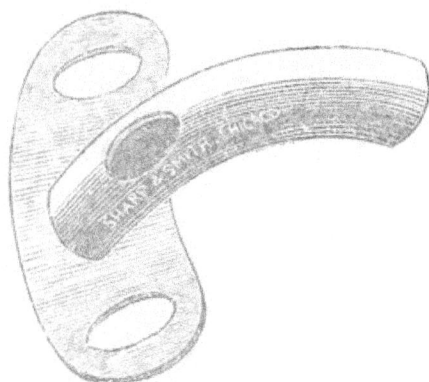

[THE TRACHAEOTOMY TUBE.]

have no tube take two fishhooks, file off the barb, then tie a piece of rubber cord to one hook and hook it into the opening, pass the cord around the neck, attach the other hook to the wound and draw the cord sufficiently tight to retain the hooks in place; this holds the wound open admitting the air. You will now apply the *mercurial blister* to the throat. After two days wash the blister off and apply a good warm poultice, using the eight-tail bandage (See Index) for a few days, giving iodide of potash for a week, and such other treatment as is generally prescribed for throat troubles. As soon · as the horse breathes easily through the nose, which is ascertained by placing the hand over the opening you will remove the tube or hooks and treat as a common wound. I have operated on a number of horses and have never seen any bad results following this operation.

EIGHT-TAIL BANDAGE.

In poulticing the throat you will always have to use the *eight-tail bandage*, which is made as follows: Take of good, stout cotton cloth enough to make a sack or pocket 12 inches wide and 17 inches long and sew three or four strings on each side of this pocket about two feet long, one-half in the center of the upper end is left open to receive the poultice, and hops steeped in vinegar, smartweed, wormwood and the *oil cake meal poultices* are to be used in warm weather, and fleece wool or cotton batting heated in the oven and sprinkled with turpentine are to be used in cold, freezing weather: When the poultice is ready you will apply it to the throat, tying the two upper strings over the neck, the two lower over the nose the remaining lower right and upper left cross between the ears, then the lower left and upper right cross between the ears. Now you will put a five ring halter on the animal to assist in keeping the poultice in place, and do not forget that *warmth* and *moisture* are the essential part of a poultice, without which there is no poultice.

ROARING AND WHISTLING.

This is considered to be an incurable disease, although some cases may recover. The symptoms are a loud whistling sound made whenever the animal is exerted either by fast driving or pulling a load.

TREATMENT.

You might try three or four applications of the *mercurial blister* on the throat extending well up to the ears and down between the jaws including considerable surface. Should you conclude to try the blister you had better give at the same time the following:

Iodide of Potash	4 ounces
Crystal Iodine	4 drams
Water	1 pint

Mix and give one tablespoonful three times a day in the feed. Should this fail your only chance of making the animal useful is by the operation called *tracheaotomy*. I have on several occasions performed this operation (using the self-retaining tube) upon animals that were useless before the operation, and today they are doing farm work with apparent ease. The tube is removed once a day, washed and immediately returned.

THE ENLARGED THYROID GLAND.

The thyroid glands are duckless glands, and when

enlarged are called *bronchocele*. They vary in size from a hickory nut to a large goose egg, and seldom interfere with the usefulness of the animal.

TREATMENT.

Lay the animal down and secure him, then make a cut lengthways of the neck long enough to allow the gland to escape, then with your thumb and fingers separate the gland from the surrounding tissue until you come to the bloodvessels which forms the attachments. These you are to tie with a silk cord, tie them twice about one inch apart and cut between the cords; be sure and tie the cord tight enough to prevent its slipping off; tie each bloodvessel separately. When you have the gland removed wash out the wound with hot water, sprinkle it over with iodiform, take a soft piece of cotton batting, sift a little iodiform over it and put it into the wound, and take two or three stitches in the skin drawing the wound together sufficient to retain the cotton. Now return the animal to the stable and give a *turpentine drench* (See Index) and a scalded bran mash. Twenty-four hours later cut the stitches and remove them, leaving the cotton to fall out of its own accord. The third day after the operation if the cotton has not fallen out you are to remove it carefully and inject a little *aqua corrosive* (See Index) once a day until well. I have never seen any bad results follow this operation. *Caution.*—Never remove but one gland at a time. Should a horse have two enlarged glands remove one and wait until he has completely recovered before removing the other.

MEDICINAL TREATMENT.

Small glands may be removed by clipping the hair off of

the enlargement and apply the following:

Saturated Solution of Iodine............... 2 ounces
Oil of Origanum.......................... 2   "
Glycerine................................ 2   "

Mix, shake and apply morning and evening for three days, then wait four days and repeat. You will now wait a month or six weeks and should there be any enlargement remaining repeat as before. Give internally one or two prescriptions of the *alterative tonic*. (See Index).

CAPPED HOCK.

This is an enlargement on the upper and back part of the hock joint *caused* sometimes by a strain, but more often by a bruise, knock, or other injury such as striking the hock against the stall or other object.

TREATMENT.

If from any known cause remove the cause for it is useless to attempt a cure without first removing the cause. Then apply the *mercurial blister*, following the directions minutely, until you have made four or five applications. (See Index). If of recent origin you might try the *white lotion* for a week or ten days before applying a blister. This will often effect a cure in cases of recent standing.

CAPPED ELBOW, OR SHOE BOIL.

These are usually the result of shoes although some large, heavy horses have them from lying on their bare feet. I have held postmortem on a few young horses that died from hypertrophy (enlargement) of the kidney, that had capped elbow, caused from the animal laying partially and then falling down on the foot. It may come from other causes.

SYMPTOMS.

It is easily detected and is an enlargement of the elbow, which is located just in front of the buckel on the belly-band. (See Skeleton for elbow).

TREATMENT.

First remove the cause. Then in the first stages bathe with hot water for several hours each day for two or three days, then apply the *mercurial blister*, following the directions. (See Index). Sometimes these enlargements are of a fibrous growth in which case they will have to be dissected out and as there is likely to be considerable bleeding I would advise you to employ a qualified veterinary surgeon. Again, scrum or pus may form in which case you are to cut it open and wash it out with warm water and inject one or two tablespoonfuls of *aqua corrosive* or *white lotion*. (See Index).

STRING HALT.

This disease is the opposite to paralysis. Most writers say that it is due to some nervous affection or nerve lesioning, but just what part they cannot tell. Some years ago I noticed an article in "Fleming's Work on Operative Surgery," recommending the division of the peroneo-prephalangeal tendon for the cure of string-halt. About this time a Mr. Price, of Findlay, Ohio, had a fine four-year old mare with a string-halt. I told him what I had read, but he was a little timid about having her operated upon, so I gave him a blister

to be applied a little above the hock on the outside of the leg
telling him that he should also apply a blister over the course
of this tendon below the hock. About six weeks later Mr. P.
turned the mare into the barnyard for the purpose of ascer-
taining whether she was any better. It was a cold, frosty
morning and the mare appeared a great deal worse but en-
joyed her freedom, running and capering about the yard
until she came in contact with a harrow that was turned up
against the fence, striking her affected leg against a tooth in
such a way as to cut this tendon off near the place recommended by Prof. Fleming. The mare was returned to the
stable that the wound might be dressed. While doing this,
however, it was noticed that the jerky action of the hock had
ceased. The next morning the case was reported to me. I
gave a bottle of *white lotion* for a dressing, the wound soon
healed and the animal was to all appearances perfectly sound.
Since then I have operated upon six cases, four of which
recovered. The operation can do no harm if it does no good,
and I think it advisable to try it in all confirmed cases. See
Fleming's Operative Surgery, page 236.

In examining a horse for soundness you should back him
out of the stall, watching him carefully. Then turn him
around quickly and he will be apt to show signs if affected.

FRACTURES—"BROKEN BONES."

They are quite common and must be treated in accordance
with their location. There are several kinds of fractures.
They are called simple, compound, commuted and complicated. A *simple* fracture is one in which the bone is broken
and the muscles and skin are not much affected. A *compound*
fracture is one in which the bones have punctured the flesh or

skin, and is hard to treat. A *commuted* fracture is one in which the bone is broken and shattered, in which case the animal had better be destroyed. A *complicated* fracture is one in which an important bloodvessel, an artery or the articulating surface of a joint is injured. Death usually follows as a result of this kind of fracture.

I shall only speak of those fractures of which I think the the chances are good for a recovery if properly cared for.

BROKEN LEGS.

Usually the best thing to do is to destroy the animal, but if you decide on treatment then the first thing to be done is to place the animal in a sling (See Page 56) and a narrow stall is preferable so that the horse cannot turn around, and he should be tied moderately short.

THE PLASTER OF PARIS BANDAGE.

Take a piece of *sheeting* three yards long and tear it in strips four inches wide, remove the ravelings and spread them at full length on a board. Now you will take some *plaster of paris* and sprinkle it over the bandage nice and evenly until they are completely covered, then you will roll them up separately until you have made four or five rolls. Lay them to one side and roll up one three inches wide without any plaster on it (a dry bandage)—lay this to one side. Now take two or three sheets of *cotton wadding*, cut in strips three inches wide and roll it up and you are ready. Stand the plaster of paris bandages on end in a half pail of water, moisten the hair on the fractured leg, smoothing it down over the fracture which we will suppose is four inches above the fetlock joint, now take the *wadding* bandages and rap them around the leg

from the knee down over the fetlock joint. Apply two layers
of the wadding tight and smooth, over this you are to rap the
dry bandage tight and perfectly smooth, now take one of the
plaster bandages and rap it over the dry bandage; rap it
tightly, but not quite as high nor quite as low as the dry ones,
as it will not do to allow the plaster bandage to come in contact with the flesh for when dry it would chafe, for this reason
the wadding and dry bandage are used. After you have
applied one plaster bandage rub in a little dry plaster and
apply the second bandage and rub it over with dry plaster.
Continue in this manner until you have applied from three to
five bandages, as the case may require. Be careful to get the
leg straight and keep it so until the plaster drys. After the
plaster is dry you must either lower the floor under the
broken leg, raise the floor under the well legs, or saw out a
piece of two-inch plank and fasten it to the well foot. This
relieves the tension on the muscles and tendons. Keep the
sling just nicely tight when the animal stands up straight, and
when he gets tired he will sag back into it and rest.

· *Keep him in the sling* six or eight weeks with a front leg,
and eight or ten weeks with a hind leg. Loosen the sling
occasionally for the purpose of currying and brushing him,
which is essential to his health and comfort.

The foregoing may be changed and diversified to suit
each particular case, bearing in mind the main points, viz:
First, a good, easy fitting sling. Second, secure the animal
so that he cannot turn around in the stall. Third, the bandages must be evenly and smoothly applied. Fourth, the
plaster bandages must not come in contact with the flesh.
Fifth, there must be at least two inches raise for extension of
the broken leg. Do not wet the leg with anything whatever
after the plaster bandage has been applied.

The food must be of a loosening nature and you may have to give an occasional dose of raw oil (one-half pint) to regulate the bowels. Should there be much constitutional disturbance give fluid extract of belladonna, one-half dram, once or twice a day and a tablespoonful of nitrate of potash once or twice a day. Should the appetite fail give quinine or iron sulphate.

POLL EVIL.

This disease is dying a natural death and the eradication of the old, low, log stables seems to have been its death knell. And I might say with the same propriety, that fistulous withers is not so common on account of our barnyards being more free from sticks, stumps and stones, and I will add another axiom: "Remove the cause and the symptoms will remove themselves. It is useless to enter into a lengthy description of these ails. Suffice it to say that a poll evil is a sore situated on the neck just back of the ears and fistulous withers is a deep-seated sore located about the neck and shoulders and there is but one way to *treat* them and that is by cutting them open and removing all of the dark unhealthy flesh, the part called pipes. These may extend to the bone and even be grown fast to the bone, in which case they must be carefully scraped loose then dress with the *aqua corrosive lotion.* There is no need of using strong remedies after you have removed the pipes with knife. Do not be afraid of bleeding, but cut a long gash with the hair and muscles six or eight inches long. I have opened them with a cut 14 inches long, the animals making good recovery, and I believe that if you cut open this class of sores and keep them open, you can effect a complete cure with the two *lotions, corrosive* and *white,* by using them alternately three days each for a few weeks or until the wound is healed from the bottom out. No wound should be allowed to close on the outside until it is closed inside. This is an important part in the treatment of wounds of all kinds and classes. For this purpose you are to introduce your fingers or hand once a day. Should you fail to keep them open this way then take a knife and enlarge the opening and keep it open at all hazards until it has filled with healthy flesh internally.

WARTS.

These are abnormal growths affecting the true skin, and are common to all domestic animals and man. They appear upon all parts of the body, legs, head, ears, eyes and mouth, and the *knife* seems to be the only specific for their removal.

TREATMENT.

[THE BULB FIRING IRON.]

First put the bulb iron in heating, then secure the animal either recumbent or standing, as the case may require; have a pail of water, sponge, needle and thread, shears and the knife at hand—you might clip the hair off around the wart and wash it before securing him. After he is secured make one cut about one quarter inch from the wart entirely around it just deep enough to divide the skin, lay the knife to one side and separate the wart from the body with your fingers; if you find any blood vessels tie them and cut them off. When you have the wart removed take the red hot iron and sear over the wound, not so much to stop the bleeding as to destroy the germs should there be any left. *After treatment.*— Bathe once a day with *aqua corrosive lotion. Do not apply a bandage or covering of any kind* as it will irritate and is a sufficient cause to produce in a healthy wound, the very thing we are trying to overcome. In fact it is very seldom that an open wound of any kind needs any protection from the air either winter or summer. My experience has led me to conclude that bandages, coverings, premature closures, and the closure of wounds with the needle have been the means not only of blemishing, but the destruction of many an animal that otherwise would have recovered.

KNUCKLING—COCK ANKLE.

This can hardly be called a disease as it rarely affects the usefulness of the animal. There may be a jerky action of the fetlock at nearly every step and a prominence in front of the fetlock without any structural change of the true joint. Knuckling is more often seen in the hind leg than the front ones. Hard driving without proper care seems to be the most exciting cause.

TREATMENT.

Feed on a loosening, laxative diet, the animal being allowed a loose box stall during treatment. Bandage the ankle and keep it wet for two or three days, then blister with the *mercurial blister*, following the directions minutely (See Index), covering the ankle all over extending a little above and below the joint. Occasionally the inflammation causes contraction and the animal is compelled to walk on the fetlock joint instead of the foot. This can be relieved by *tenotomy*, a division of the tendons between the hock or knee and fetlock joint, and is usually followed by favorable results if properly performed.

CHOKE IN THE HORSE.

Some horses are hoggish eaters, and especially when fed in a high manger are apt to get choked on oats. This may be prevented either by placing the feed box on the floor, which is perhaps the best way, or by putting a few round stones the size of a goose egg in the feed box. Should a horse get choked you are to pour a little raw oil down the throat, carefully rubbing the choked place with your hand. Jumping the horse quickly over some object 18 to 24 inches high will often remove a choke. Never try to force a stick or piece of wood of any kind down a horse's throat.

CHOKE IN THE COW.

Cattle are quite frequently choked on apples, potatoes, turnips, etc.

First procedure.—With your hand rub and manipulate the throat working the object up as near the mouth as possible, and have an assistant hold it there. Now put a *clevis* or some object in the mouth to hold it open and run your hand down into the mouth and get the choke. Should you fail in this adopt the

Second procedure.—Take a piece of smooth wire 10 feet

long about the size of a pail bail, double it in the center forming a loop, leave the loop a little rounded and about four
inches wide; oil this and introduce it into the mouth. Have
the choke held by an assistant; now carefully work the wire
past the choke when it will come between the two wires,
which you can feel with your hand, now carefully withdraw
the wire, object and all. This operation has always been successful with me.

Do not try to force a choke down into the stomach until
the last resort, and then use the doubled wire a ball attached.
The ball should be one and a half inches in diameter, perfectly smooth and a little flat or hollow on the side that goes
goes down against the choke. A piece of inch or inch and a
quarter rubber hose will answer very well for forcing a choke
down into the stomach, but a bar, pitchfork, broom-handle, or
whip stock will kill more than they will cure.

ANIMAL CASTRATION.

CHAPTER IV.

THE EFFECT OF CASTRATION.

This particular piece of surgery is of all others the one most often performed upon the domesticated animal, and is followed by certain peculiar effects. The most striking change produced by the removal of the testicles is manifest in the character and disposition of the animal, which becomes at once, in a double sense, an *altered* being submissive and docile willing to become the obedient and useful servant of man. Thus we find the vicious stallion, the unmanagable bull, the dangerous boar, the hysteric mare and kicking cow transformed into the useful gelding, the gentle ox, the fattened barrow, the quiet, submissive mare and the productive cow, as the result of this operation. The male animal assumes the

character and form of the female not only in appearance but
in voice, which loses its resonant sound of the stallion. A
like change takes place in all other male animals and man to
acquire a resemblance to the female as an effect of castration.
The altered bull has a weak, feminine voice, his horns are
longer and more curved; he has exchanged his wild and
threatening aspect for a mild and comely visage. Again the
castrated animal ceases to exist as one of a species, but main-
tains an individual life in which the food he absorbs instead
of being in part appropriated to the office of reproduction of
his kind, is all devoted to his own individuality. When the
food is in excess of the amount required for the support of the
animal it follows that the surplus of the nutritive material
becomes stored in the connective tissue and muscular structure, the flesh assuming a more nutritious and juicy quality
than that of the uncastrated animal. At the same time it
loses the peculiar odor and testicular taint of the entire
animal.

THE AGE TO CASTRATE.

The question is frequently asked, "Of what age should an
animal be castrated?" We reply, at any age, though the
younger the animal is castrated the better, since a young animal is not as liable to unfavorable complications following
this operation as an old one. With ordinary care an unfavorable result in a sucking animal is very rare, indeed almost
unheard of. This will apply to all male animals. The age
at which a colt should be castrated depends much upon the
object in view. If you are breeding for early profit geld at
age of three months, for three reasons. *First.*—Colts gelded
at this age are cut up in the throat-latch nicer and square out
in the hips much wider, giving them more the appearance of

the mare. *Second.*—They will be ready for market a year sooner. *Third.*—They heal readily and the chances for recovery are all in your favor.

However, a colt gelded at from two to three years of age is tougher and more rugged, though a little "staggy" about the neck and a little "peaked" in the hips, and therefore is less valuable in market. All things considered, the best age to castrate a colt is when he is about one and a half years old.

SEASON.

Most, if not all writers on this subject recommend that castrating be performed either during the spring or fall months. I am, however obliged to differ, with them, giving the preference to the months of June, July, August and September since a large experience in this class of work at various periods of the year has proven to me conclusively that horses castrated during these months do better and escape many of the otherwise unpleasant after effects such as swelling of the sheath, colic, etc. This I attribute, first, to their general health being better after a run at grass; second, the ground is warm and dry and they do not take cold while lying down; third, the flies keep them in constant motion, which not only prevents swelling by keeping the absorbents at work, but also keeps up a free discharge from the wound— a very essential thing in an operation of this kind. Horses castrated in hot weather recover in from two to three weeks, while those castrated in the early spring months take from six weeks to six months to get well; some never recover.

PREPARATION.

When a horse is to be castrated the following preparation is essentially necessary, *regardless of age*: From 10 to 14 days before castrating take a pail of warm soap suds and thoroughly wash out the sheath, examining the end of the penis carefully for a "bean" which is sometimes imprisoned there, and which can be detected by pressing carefully on the organ, as it feels hard to the touch. This must always be removed. After washing out the sheath take

Glycerine.................................. 1 ounce
Sweet Oil.................................. 1   "
Carbolic Acid....... ....15 drops

Mix well together and anoint the sheath thoroughly, applying it well back around the penis. You will now, *if a matured horse* give a laxative ball followed by a scalded bran mash; *if a colt* omit the ball and give the mash. Then send word to your intended operator that you shall expect to have your animal castrated in from 10 to 15 days.*

CRYPTORCHIDE (RIDGLING) CASTRATION.

This engraving shows the location of the testicle in the cryptorchide. These horses are commonly called original horses. And for the benefit of those wishing to post themselves I have defined the three common terms used to designate horses of this class.

Prof. Liautard, of the New York Veterinary College, in his treatise on animal castration has adopted the name cryptorchide, which is more expressive of the abnormal conditions; for this reason we have used it. The name Ridgling

---

* NOTE—Do not put off this washing for the operator to do, as the washing of itself is liable to cause some irritation and swelling, which will subside in a week or ten days if the foregoing instructions are minutely followed. The parts, not only the parts operated upon, but the whole animal should be in prime condition. If the animal has or appears to be bordering on disease, you must put him under treatment, deferring the operation for a time.

comes from the French word *Ridge*, meaning *halved*. Thus
it will be seen that a ridgling horse is a *half* castrated horse,

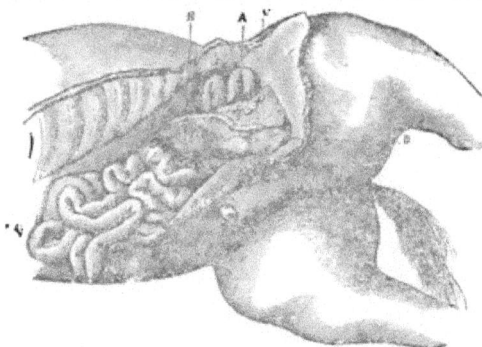

[THE CRYPTORCHIDE—" RIDGLING."]
A. Vasdeferens; B, Floating Testicle; C, Epididymis; D, Spermatic Cord.

regardless of the location of the testicle. "Original,' 'The
first of its kind.'" Hence it will be seen that the primitive
meaning of the word original could not possibly be con-
strued as relating to a thing which has been remodeled or
fixed over. *Cryptorchide*, "Having a secret testicle." This
name is the proper one, as it carries with it a meaning. However, the name cuts no figure in the operation, and we feel
that this little work would be quite incomplete without a word
on this important branch of veterinary surgery—a branch
which seems to baffle the skill of so many good vets. Yet, it
is simple and easy when once you know how. During the
months of July and August, 1889, I castrated 53 cryptorchides
in Michigan and Northern Indiana, with perfect recovery in
every instance, and I have castrated over 300 cryptorchides
in the last five years without a death, which is ample proof of
my ability, and I will go to any part of the United States or
Canada and teach any qualified surgeon how to operate suc-
cessfully for $100 and expenses. This is the only secret in

surgery that I have withheld from the public in the publication of this little work.   References furnished on application:

Leonard L. Conkey.

THROWING HORSES FOR CASTRATING.

The horse should always be laid down to insure perfect safety to the operator and those who assist him.   Some recommend castrating while standing, but it is not the proper way. I have seen men who claimed to be experts attempt to castrate a stallion while standing.   The operator would get one testicle out, or perhaps only an incision, when the animal would become so enraged that he would have to be taken out and thrown before they could complete the operation.   Again the testicles of some colts are retained in the inquinal canal, and do not descend into the scrotum until they are two or three years old and upwards.   These are usually called

FLANKERS.

And it is next thing to an impossibility to castrate such horses standing.   Furthermore, I am sure that no living man can castrate an *abdominal ridgling* while standing.   In view of these facts, and others too numerous to mention, it appears to me to be the most rational way to lay a horse down at once, which can be done on a blanket with *Conkey's Securing Harness and Hobble.*   However, there are many different methods of throwing horses, all of which have more or less merit.

FARMER MILES' METHOD.

[FIG. 34.]

The accompanying engraving are reproduced from Prof. Fleming's New Operative Surgery, Pages 36 and 39. Fig. 34 represents Farmer Miles' mode of applying ropes for

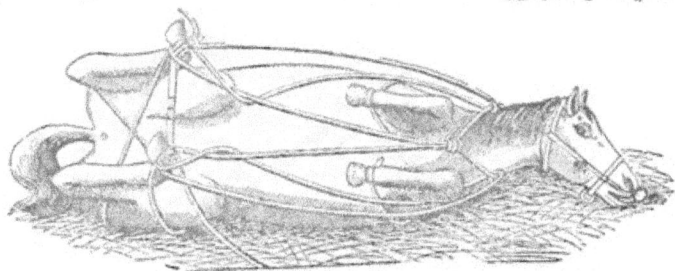

[FIG. 36.]

throwing horses. Fig. 36 represents the horse thrown and secured. For lack of space we omit the lengthy description, mode of applying and workings of Miles' method. Suffice it to quote that "with practice an animal may be prepared, cast, tied up and released again in from 10 to 12 minutes. Five active men are sufficient to assist."

L. L. CONKEY'S METHOD.

FIG. I.—THE HORSE PREPARED FOR CASTING.

Figures 1, 2, 3 and the rope A represent the Conkey Patent Hobble applied; Figures 4, 5, 6, 7 and 8 show the Conkey Securing Harness applied; A, the operator; B, C and D, assistants; E, the Conkey Operating Hood.

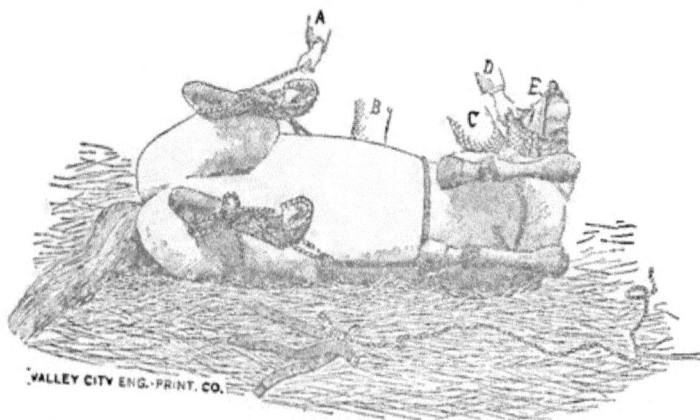

[FIG. 2.—THE HORSE SECURED AND THE HOBBLES REMOVED.]

A, right hand; D, left hand; B, right leg; C, left leg of the assistant who holds the head.

With this harness and hobble I can lay (not throw) a horse, regardless of size and disposition, on a horse blanket, not once in a while, but every time, with only two good assistants, and more than three are only in the way.

The following will serve to give an idea of the length of time required to prepare and secure an animal. In May, '88 I castrated six straight colts and one abdominal ridgling in 40 minutes without apparent haste. This work was done at Mr. Harm DeLong's, Penn, Cass Co., Mich. Mr. J. Garwood owned the ridgling, and there were the Bonine brothers and at least a half dozen other farmers present. Each animal was led from the stable to an adjacent field and returned to the stable singly after being operated upon. The following is the outgrowth of an assertion made while preparing to cast the horse represented in Figures 1 and 2, which explains itself:

GRAND RAPIDS, MICH., DEC. 20, 1889.

It is with pleasure I acknowledge that Dr. Conkey's Hobbles and Harness, for securing a horse in the recumbent position, is certainly the best, easiest and quickest I have ever seen. Dr. Conkey, with the aid of two assistants, claimed he could lay a horse on a horse blanket with his hobbles and did so in presence of Mr. A. N. Albee, liveryman; Mr. Reed, of the Valley City Engraving and Printing Co., myself and several others. The horse was a large, powerful animal, but the task was as easy as though he was but a sucking colt.

Yours respectfully,

P. H. O'BRIEN.

Dr. Wm. Rose, V. S., graduate of the Toronto Veterinary College favored us with the following, which speaks for itself:

DR. L. L. CONKEY:

Dear Sir---In answer to your inquiry about the Securing Harness and Hobbles bought of you will say that I would not be without them for many times their cost. An animal can be laid down easier, quicker, secured perfectly and safely. Besides this it can be done in less space than by anyother method I have ever seen. And for the benefit of the profession will say that Conkey's method is not only admirably adapted to castrating, but also to any and all conceivable operations, neurotomy, miotomy, peroneo-prephalangeal, tenotomy, firing, shoeing, etc. No progressive veterinary surgeon can afford to be without your Hood, Harness and Hobbles.

Yours with respect,
WILLIAM ROSE, V. S.
128 East Fulton St., Grand Rapids, Mich.

[THE CONKEY PATENT.]

L. L. Conkey's Patent Automatic Chain Buckle, Patented Jan. 1889, in the United States and Canada. For price of Hobbles, Hood and Securing Harness address Leonard L. Conkey, Grand Rapids, Mich. Full instructions and illustrations sent with every outfit, which are guaranteed as represented or money refunded.

ACCIDENTS INCIDENT TO THROWING HORSES.

The accidents liable to occur in throwing horses down for operations, and while down are numerous, and may occur, the operator having adopted all due and proper precautionary measures.

I find on page 48 of Prof. Fleming's Operative Surgery the following report of accidental fracture of the back. " Of my 14 cases 13 occured while the horse was on its side, the 14th while on its back. My cases occurred with the following number in various kinds of horses. One was a heavy dray-horse, two English thoroughbreds, seven half-bred blooded stock, and four were of the blooded country stock. Age appeared to make no difference. The youngest horse, a thoroughbred English stallion was three. The oldest, a coach-horse (stallion), was 21 years old. It is understood, says the AUTHOR, that the character of the operation itself has no influence whatever on the causation of vertebral fractures; but to be exhaustive I will mention that one case occurred in castration, one case in operation for fistula, two cases in extraction of molar, four cases in neurotomy, four cases in firing, two cases in spaying." FLEMING."

In my own practice I have never met with the accident of *fracture* in any form; neither have I met with an accident of any kind except in one instance, in which I threw a large, iron grey two-year-old. This animal was as nice a colt as I ever roped with the exception of his feet, which were badly cracked or split, and I threw him for the purpose of fixing them. This I was compelled to do on account of his vicious disposition. A few days after one front leg was found somewhat swollen, and about two weeks later one hip was noticed to be shrinking away. The colt had fought incessantly while down, straining himself sufficient to produce the foregoing results. Thus it will be seen that accidents are at times unavoidable, and no man living dare say that Prof. Fleming is a careless operator, or that he was in anyway to blame for the accidents reported. Yet I am of the opinion that if he and others would adopt the mode of securing horses illustrated on Page 149, they would never again hear of a broken back. I have cast several thousand horses in the last 10 years for different operations, and with the one exception they have all gotten up without assistance or any apparent bad result. Therefore I think I am justified in saying, lay the horse down for castrating, do it carefully secure him firmly and you are ready for the knife.

THE CASTRATING KNIFE.

Take the knife in your right hand, grasp the point of the sheath with your left, the fingers entering the sheath; take a good, firm hold drawing the sheath forward (toward the horse's head) and upward from the body as the horse lays on his back, holding it firmly. You will make two incisions (cuts) about four or five inches long and about one and one-half inch from the middle line of the sheath, running the same way parallel with the sheath through the scrotum. Make the first two cuts through the skin on each side, then cut through the fibrous tissue. Now let go the sheath and force the testicles (which will have a white covering called the tunica) through the opening, holding it firmly between your thumb and finger, make a free cut with the knife through this white

covering when the testicle will be exposed. Should the testicles be large and well developed, as is often the case, it will
not be necessary to take hold of the sheath, but simply grasp
the lower testicle between thumb and fingers with a strong
hold making one sweeping cut with the knife exposing the
testicle at one or two cuts. Take hold of the testicle with the
left hand draw it out gently and should the white covering
form a sack like cover around the cord, as is often the case,
you are to introduce the knife carefully by the side of the
cord and cut open this sack down toward the horse's body as
far as you can handily and take the ceraseur in your right

[THE ECRASEUR.]

hand and pass the chain over the testicle well down to the
body around both the spematic cord and a part of the "white
covering," now take up the slack of your chain as quickly as
possible. Look now and see that there is nothing between
the chain and cord, the white covering being next to the instrument; now turn the instrument until you feel that the
chain is very tight or commencing to crush, then stop long
enough to count six, make a half turn, count six, another half
turn, count six; continue this turning and counting until the
chain has been drawn into, not through, the loop of the
instrument, this you can tell by the instrument turning more
easily and by the feeling at the end. You will now grasp the testicle with the left hand and pull it off, holding the instrument
well up against the horse with the right hand while you pull.

You will now remove the other testicle in the same manner. If the testicles are small you may, at your pleasure, remove them both at once; but you must be very careful and not get the scrotum or sheath in the instrument or you will be liable to have bleeding following the operation. After removing the testicles you will now bathe the wound with cold water, or with carbolized oil (See Index). You will now remove your ropes as quickly and carefully as possible and let the horse up and should there be

BLEEDING AFTER CASTRATION

sufficient to cause any uneasiness, pour cold water over the hips and loins. Should the bleeding continue after a few minutes take a lump of alum about the size of a hen's egg, pound it fine, then pour on one pint of boiling water, add cold water to cool, then inject with a syringe half or two-thirds of this into the rectum. Then saturate a sponge, ball of cotton, or old cotton cloth with the alum water and force it up into the wound allowing it to remain undisturbed for 24 to 36 hours, and then be removed very carefully. This treatment is successful except in rare instances, when the animal will have to be thrown and the *artery* tied or you may apply

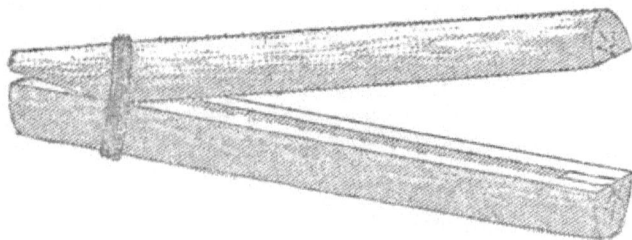

CONKEY'S CASTRATING CLAMP

which is best made of seasoned bob shumake, sweet alder or

paw-paw on account of its being light and having a pith, which is easily removed, forming a cavity to receive the caustic medicines which *must* be used.   The clamps are about five inches long and from three-quarters to one inch in diameter, with a groove cut around the end to receive the string used in tying them together.   They are split in halves, the peth removed and one piece beveled from the notch or groove to the end as shown in the engraving.   Just before laying the horse down you will take fresh lard, tallow, butter or cosmoline and grease the clamp over filling the groove and rubbing it all over the flat surface of the clamp; this prevents the clamp from sticking to the cord when removing it, as well as holding the

CLAMP POWDER.

All clamps used for castrating must be medicated before they are used, and for this purpose there is nothing better than the following:

Finely Powdered Corrosive Sublimate	....	1	ounce
"	"	Red Percipitate	 1	"
"	"	Willow Carbon	 1	dram

This powder is to be thoroughly mixed and sprinkled pretty freely over the whole flat or inside surface of the clamp just before it is applied to the cord.

CASTRATING WITH THE CLAMP.

Lay the horse down, make your incisions, and take the testicles out the same as for the ecraseur, put the clap on from forward back close to the body, clamping in the "white covering" and all the loose cord, (remove the lower testicle first for convenience) close the clamp with the

CLAMP TONGS

while tying it, and be sure to tie the clamp well that it may not slip off, next you are to cut the testicle off about one-fourth of an inch below the clamp rubbing the severed end of the cord over with the clamp powder. Now pour some carbolized oil (See Index) around the clamp working it down into the wound. Now remove your ropes quickly, but carefully and let him up. Tie him in a good stall where he cannot lay down for an hour or two until he gets quiet, as they are frequently very restless, stamping and sweating profusely, then turn him out into the pasture. The after care will depend much upon the various complications which are liable to follow. But if you have your horses gelded in hot weather they will get well without even swelling.

THE CARE OF A HORSE AFTER CASTRATING.

Tie the horse in a comfortable stall for a few hours, or until he is perfectly quiet, then give him a turpentine drench (See Index) followed by a *scalded* bran mash, then you will give one-fourth pound of epsom salt every third day until you have given one pound. Feed on a laxative, nutritious diet, scalded (not raw) bran, oats, etc. Give him all the water he will drink. Should he swell much give him one-half pint of raw linseed oil, containing one or two ounces of sweet spirits of nitre once a day until the swelling disappears, continuing the other treatment as before, unless the bowels should get too loose, in which case you are to discontinue the salt until the bowels set and are perfectly natural.

Open the wound two or three times a day for a week by inserting the hand into the wound the whole length, (not the tips of the fingers, but the whole hand), give gentle exercise at first, then *trot* a mile or until the swelling is gone and the parts are perfectly natural. If the horse is old enough to work it is best to work him a little every day; if not he may be tied beside your work team and driven half of each day until well. *Never allow a horse out in a cold rain for at least three weeks after castration.* Never go to bed leaving any swelling in the privates of a colt, but jog him about until it disappears. *Exercise will remove swelling.* Never allow a man to operate who hesitates, gives evasive answers, guesses or thinks he can do the work, etc. Remember that a horse is sound that has both testicles, regardless of their location, and that the removal of one testis makes him unsound and an unlawful horse to sell.

THE BULL.

The stanchions that are in common use for stabling cattle is about as good a place as you can put a large bull, say one year old, to castrate him. Put him in the stanchion, tie a cord to the left hind leg below the dew claws, draw the leg forward a little and make the rope fast to the horns or stanchion, then grasp the bag in your left hand with a firm hold, pull down hard enough to stretch the skin, which you will cut open the whole length of the testicle on the side (be careful not to cut the testicle too much, as it interferes with the operation). As the testicle slips out of the white covering you are to separate the testicle from the white covering (to which you will find it attached at the lower end of the testicle) with your thumb and fingers, then carefully draw the testis down with one hand pushing the covering up into the groin with the other. Keep pushing and pulling until you have two or three feet of cord or even more. It will do no harm to pull until the cord breaks off, but if it refuses to break when you have pulled it out until it is no larger than a small rye straw and perfectly white you may scrape it in two with the back of your knife. Remove the other in the same manner. Take hold of the lower point of the scrotum (bag) and pull it down straight and fill the cavity with common salt and the operation is complete.

YOUNG CALVES

are to be laid on the side and operated upon in the same manner. Do not cut off the end of the bag but slit it open on the side and draw the testicles out until they break off, straighten the bag and put in a little salt and let him go. Should a bull swell in the bag you are to open the wound and allow the pus to escape.

RIDGLING BULLS.

Occasionally you will come across a bull that has but one testicle in the scrotum, and you are at a loss to know where the other is. In such cases you are to feel all over the belly. I have found them beside the penis near its end, down the thigh, deep in the flank and other locations, but never inside of the abdominal wall ; so if you look carefully you will find it somewhere, and wherever you find it make a good, large opening that the pus or matter may escape freely and draw it out until the cord breaks off Sprinkle a little common salt in the wound after its removal.

THE BOAR.

The boar must be laid on the left side held by an assistant with his right knee on the neck, the right hind foot in his right hand, and the right front foot in his left hand (let him squeal). The operator now places his left knee on the hog's flank, grasps the lower testicle in his left hand between the thumb and forefinger tightly, (be sure that the thumb covers the middle line between the testicles ; this prevents you from cutting across it, an accident which causes much swelling), make a long sweeping cut with the knife two and a half or three inches long and pretty low down so that there will be a free discharge. As you make the cut the testicle will come out in such a way that your thumb and finger will meet between the testis and white covering, which is to be torn loose ; pull the testis until the cord breaks off a foot or so long. Do not cut the cord off, but pull it out, and if you make a good, large opening you will have no trouble with them.

IN HOT WEATHER

Take of pine tar and spirits turpentine equal parts and warm it together, tie a rag on the end of a stick for a swab and rub the mixture onto and into the wound. This will prevent maggots from getting in.

RIDGLING BOAR.

Not often but occasionally you will find a boar with but one testicle in the scrotum. Should this be the case you are to feel carefully in the groin, between and in front of the hind legs. Should you fail to find it in this locality you are to lay the hog down with the side up corresponding to the missing testicle, have the upper hind leg drawn back by an assistant. Cut the bristles off from a large spot in the flank, then make an incision the same as for spaying a sow (See Index). You may have to make the incision large enough to admit the whole hand which must be done very carefully, feeling for the testicle between the kidney and the bladder. When you have found it remove it with the *spaying shears*, or tie a linen thread around the cord close to the body and cut off the testicle about one-half inch from the string; allow the cord and string to fall back into the abdominal cavity and sew up the opening with cotton twine. Sew it "over and over" with the stitches not more than half an inch apart. Pass the needle down through the skin, flesh and thin, white lining of the belly, then up through the thin, white lining, flesh and skin, and so on until you have closed the opening. Rub on a little tar and the operation is complete.

SCROTAL HERNIA IN THE HOG.

THE RUPTURED BOAR.

Is castrated by making a long cut through the skin, dissecting out the white sack that covers the testicles and bowels, the same as the stallion, but instead of using a clamp you will tie a stout string around the sack close to the body, and it is

a good practice to have a needle on the string to take a stitch
or two through the sack to prevent the string from slipping
off; cut the sack and testicle off an inch from the string;
leave the string long enough to hang out of the wound.   This,
if properly done, will make your hog perfectly smooth.

THE RAM OR BUCK SHEEP.

The old ram often dies from castration by the methods
now in vogue, which I will not stop to enumerate, but will
endeavor to make plain a method which has always (without a
single exception to my knowledge) proved successful.

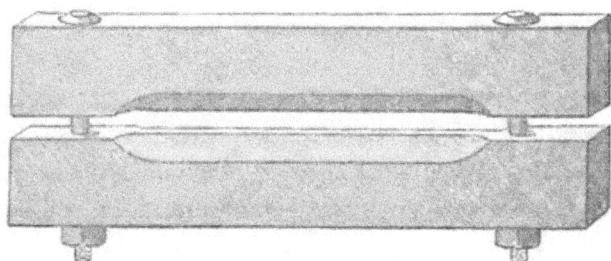

[CONKEY'S BUCK CLAMP.]

Take two pieces of dry oak wood five inches long and three-
fourths of an inch sqare, lay them together in a vice and bore
a one-quarter inch hole through both ends of each piece three-
quarters of an inch from the end for bolts, now you will bevel
the face side (that is the sides that touch together) commencing
one inch from the end and bevel towards the center (leaving
the ends square where the bolts go through); you will bevel
each corner off until the surface will be a half inch or less
in thickness.   Now smooth off the outside corners and it is
ready for use.   Lay the ram on his back, have an assistant
take hold of the bag and testicles and draw them up away

from the body as far as he can and hold it there while you apply the clamp; put it on lengthwise of, and close to the body as you can get it. One bolt is to be put in the clamp before it is applied, you will now put in the other bolt and proceed to tighten the burs with a wrench as tight as you can get them handily, then cut the bag, testicles and all off about one-half inch from the clamp and let him go. The clamp must be allowed to slough off. By this method there is no possible way for septic germs to enter the body from the fact that nothing can pass between the clamp. I have never heard one word of complaint where this method was adopted. No after treatment required.

THE LAMB.

This is a very simple operation, and usually successful in its termination. Stand a barrel on the barn floor in a convenient place, have a can of pine tar and spirits turpentine, equal parts, warmed together, a small brush or swab, a sharp knife, and you are ready. Your assistant will now bring a lamb, lay it on the barrel with both right legs in his right hand, and both left legs in his left hand, front legs outside, feet sticking back and the hind legs inside, feet along the breast. You now take hold of the end of the bag and cut about one-third of it off, then take hold of the bag with the thumb and finger and force the testicles out taking hold of them with your teeth and pulling them out until they break off. I recommend extracting them with the teeth for two reasons: First, it is much the quickest way. Second, you do not get the oil or grease from the wool into the wound, which seems to set up an irritation, which you are sure to do if you remove them with your fingers. Rub a little tar on the

wound and let him go. Use the tar quite freely as it will
keep the blow fly away.

SCROTAL HERNIA IN OLD HORSES.

BREACH, RUPTURE INTO THE SCROTUM.

When it becomes necessary to castrate a horse with
scrotal hernia, secure him in the same way as for ordinary
castration. Have a good, strong clamp previously prepared
(it is a good practice to secure the clamps with quarter inch
iron bolts for a large horse) cut through the scrotum carefully,
but do not cut the tunica vaginalis (white covering). Make
a good, long cut through the scrotum and dissect out the
white covering by skinning back the scrotum until the white
cover or sack that holds the bowel and testicle is all free as far as
the body. Then you will force the bowell back into the body
retaining the testicle. Now have an assistant apply the clamp
moderately tight close to the body, then make a small opening into the sack freeing the testicle. Draw it and the fleshy
part of the cord out, fasten the clamp perfectly tight, cut off
the testicle and sack about one-half inch from the clamp,
take a needle and sew up the sack below and the whole length
of the clamp; sew over and over. This forms a seam preventing the clamp from slipping off. The clamp must not be
removed, but allowed to slough off, which will take from 7 to
14 days. In bad cases you may have to operate two or three
times before the animal is perfectly smooth. This operation
has always proved successful with me. *Internal Treatment.*—
Give a laxative ball or drench followed by a scalded bran
mash once a day with an occasional dose of saltpetre, say a
tablespoonful once or twice a day for a few days.

SCROTAL HERNIA.

RUPTURE, OR BREACH IN THE COLT.

If the colt seems to suffer no inconvenience it is not best to interfere with him, as a spontaneous cure is often affected when the colt is put on dry feed, which causes the large intestines to drop down on the floor of the abdomen crowding the small ones forward out of the scrotum.   However, should the bowel become strangulated causing colic pains, he had best be castrated at once, which is done in the same manner as in old horses (See Index).   If properly done there is no danger to be anticipated.

THE DOG.

Tie a stout string around the nose and back over the head to prevent him from biting, and operate the same as with the boar. No after treatment is required.

THE CAT.

Take a two-bushel bag, spread it out upon the floor, lay the cat down carefully with all the legs extended toward the head and the scrotum at the edge of the bag, then roll the cat up in the bag as tightly as possible, then set down in a chair and lay the cat between your legs with the cat's head in the chair and under your leg and the hind parts up between the legs with the back toward you and castrate the same as a hog.   No after treatment required.

SPAYING THE MARE.

This operation is very rarely performed and never except in cases of bad kickers. A worthless kicker will become kind and gentle after this operation, which is accomplished by cutting up through the vagina and drawing the ovaries down through the opening where they are to be removed with the long ecraseur. None but an expert should be allowed to undertake this operation, and then it may prove fatal.

AFTER TREATMENT.

See internal treatment of the horse after castration. Inject the vagina once a day with the following:

Warm water............................... 1 gallon
Corrosive Sublimate .... 1 dram

SPAYING THE COW.

This operation is usually performed six or eight weeks after calving, and is best done while standing by cutting a hole up through the walls of the vagina close to the mouth of the womb large enough to introduce two fingers, with which the ovaries are pulled down and removed with the long ecraseur.

SPAYING THE HEIFER.

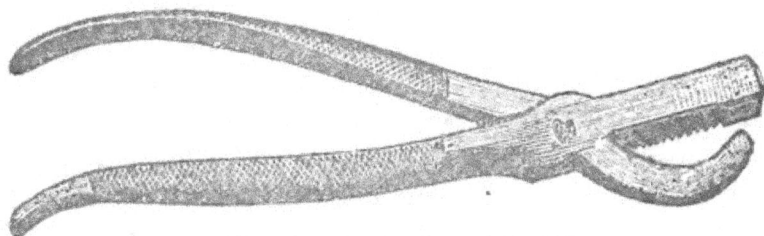

[THE CONKEY SPAYING INSTRUMENT.]

After fasting twelve hours clip the hair all off from the angle of the right flank between the last rib and the point of the hip, remove the dirt with a stiff horse brush, then lay her down on the left side using four hobbles. After she is down remove the hobble from the right leg and have it drawn back by an assistant. Make an incision up and down large enough to admit the hand, feel for the horn of the womb and follow it until you find the ovaries; feel for the lower one first and remove it with the Conkey Spaying Instrument, then remove the other ovary in the same manner. Return the parts and sew up the flesh first with *carbolized cat gut,* sewing over and over, then sew up the skin with carbolized silk or anything you like; tie each stitch separate and from one-half to three-fourths of an inch apart. Select warm, dry weather, and great care must be taken to prevent any dirt, hair or other foreign substance from falling into the abdominal cavity. Dress once a day with

Corrosive Sublimate........................ 1 dram
Water.................................... 1 quart

SPAYING THE SOW.

The sow is also to be fasted for twelve hours, and perhaps this rule would hold good in all animals that are to be operated upon.

The instruments used are a piece of cord (clothesline) three or four feet long, a pail of water and sponge, a sharp razor, a knife and curved needle three inches long. The knife should be a handle with a blade two and a half inches long by one-half or five-eighths of an inch wide, spear-pointed and very sharp. Take a two-inch plank one foot wide by 10 feet long, lay one end on the ground the other on the fence about three feet from the ground, drive a large nail or spike into the upper end of the plank, tie the ends of the cord together forming a loop, hang the loop over the spike in the plank, set a barrel on the opposite side of the fence, set the pail of water on this barrel, lay your instruments beside it and you are ready. Stand with your left hand to the fence when facing the plank. Your assistant will now bring a hog, having hold of both left legs with the back towards him, he hands you the left hind leg which you take in your left hand, and he takes hold of the right front leg dropping the hog on the plank well up to the fence back toward you. With your right hand you lay the doubled or looped cord between the hind legs passing your right hand through the loop taking hold of both hind feet with your right hand; the cord is now between the hind legs and over your right arm. Now take the cord in your left hand and drop it back around the hind legs; this frees your right hand and at the same time forms a half hitch around both legs of the hog at once. Do not let go of the feet with your right hand until the hog has been drawn down the plank far enough to tighten the cord. Now take the sponge and wet the flank (not too wet), take the razor and shave off the bristles just in front and below the point of the hip, place the thumb of your left hand on the point of the hip and made a span with the forefinger towards the belly and a little forward; move the thumb up to the finger pinching up the skin in a fold, which you are to hold tightly, take the spaying knife and make a bold, free cut across this fold and let go the skin; now you have a cut an inch and a half long running crosswise of the body. Next take hold of the knife something as you would a pen to write, with the cutting edge lengthwise of the hog and across the cut already made. As the hog makes an expiratory movement (that is to say as the air leaves the lungs) plunge the knife down into the hog about the middle of the cut already made, withdrawing the

knife quickly, lay it on the barrel and insert the first finger
of the right hand feel for the ovary or the horn of the uterus.
You may have some little trouble in finding this at first, and
I would advise you to use an *uterine sound*, which is made in
the shape of a letter **S** and about a foot long. This you
will oil and introduce through the natural opening into
the vagina with the left hand lifting up the organ so that you
can readily feel it with the finger. When you get one horn
you are to overhaul it until you come to the ovary, which has
a reddish looking soft part attached called the *fimbriated body*.
This you are to tear off with the ovary, for if left the sow is
likely to come in heat again. After you have taken off the
ovary from one side you are to overhaul the uterus until you come
to the fork where the two horns meet; be very careful not to
break the uterus off. When you get hold of the lower horn
you are to overhaul it in a like manner until you find the
ovary, which is to be twisted and broken off. Do not cut
them off, as that would cause a great beal of bleeding   Return
the uterus carefully; be sure that you get it through the thin,
white lining of the belly.   Take the needle, previously threaded,

pass it *down* through the *skin* at the middle of the skin cut,
then *down* through the *flesh* and thin, *white, lining* of the belly
in the middle of the flesh cut, now *up* through the *white lining*
and *flesh* on the opposite side of the flesh cut, and lastly up
through the skin tying the string firmly.   Daub on a little tar
and let her go. In spaying large hogs the uterus should be
returned as fast as it is overhauled, never allowing more than

a foot exposed to the air at once, as it is liable to swell and give you a good deal of trouble in returning it. Old sows may be spayed with perfect safety, but the best age is from 8 to 12 weeks old. With a little practice you can spay a sow in about three minutes. A year or so ago I handled for Mr. Dennis Dailey, of Dowagiac, Mich., eighty hogs (averaging about a hundred pounds each) in three hours, the majority (57) where sows, which I spayed, castrated the boars and punched a hole in each ear of every hog. Mr. F. Foster kept tally, Mr. Daily handed me the instruments while his son and hired man caught the hogs. Eighty hogs in 183 minutes—a 2:15 gate—without a skip (death), and I should be pleased to hear from the man that can break the record.

SPAYING THE BITCH.

This is quite a difficult operation to perform, as it is hard to distinguish the difference, by the feel, between the fat and the generative organs. For this reason it is always best to provide yourself with a uterine sound, and the common male catheter, which can be bought at any drug store for 25 cents, answers well for this purpose. *Modus Operandi.*—First take a piece of wool twine or other stout string, tie it around the nose and under jaw tightly, then up over the back or top of the head so that it cannot slip off (See Engraving); this is to prevent them from biting. Now lay her on a table or other convenient place with one man at each end of the dog holding the legs with her back towards you. Make the cut on the median line in front of the two last teats and between the next two, two or three inches in length and lengthwise of the belly following the

light colored mark in the middle. After you have cut through introduce one or two fingers of the right hand through the wound and the sound or probe from behind into the vagina; be careful and do not get the probe into the bladder, but work it carefully into the vagina. When you can feel it with the fingers of the right hand this tells you when you have the horn of the uterus, which is to be carefully drawn out, when you will take hold of the uterus (pup bed) with the left hand, pull enough to make it tight, then follow it with the forefinger of the right hand in toward the back until you find the ovary, which must be torn loose with the fingers and brought out. You must be careful or you will break off the uterus and leave the ovary, in which case she will come in heat, but will not breed. Tear the little cord-like uterus loose from the fat, returning the fat. Now follow the uterus back until you find the fork where they unite, get the other horn and remove it in the same manner. When you have them both out twist them off, or, what is better, remove them with the spaying shears at the point where the two horns meet. Return the fat and sew up the wound by sewing it over and over, down through the skin and flesh, up through the flesh and skin with three or four stitches. Now cut the stitches and tie them separately, which completes the operation. Give at once internally one-half ounce of syrup of buck thorn. About the third or fourth day cut the stitches at the side of the knot and the dog will remove them herself. Results always favorable in my practice.

COMPLICATIONS FOLLOWING CASTRATION.

While the operation of castration is comparatively simple and easy, yet it is not entirely free from accidents or complications. Indeed there are some things of a very serious character which develope themselves regardless of the skill or care with which the operation may have been performed, such as *colic, bleeding, swelling of the scrotum, gangrene, abscesses, locked jaw, etc.*

COLIC.

This is liable to appear shortly after the operation.

SYMPTOMS.

Pawing, stamping, lying down, rolling, etc.

TREATMENT.

Give one-half ounce of chloral hydrate in a teacupful of water as a drench, and repeat the dose every half hour until relieved, then you are to give from one-half to one pint of raw linseed oil at once. Treatment is always satisfactory.

ABSCESSES.

These result from the wound closing too soon after the operation, which imprisons the pus or matter. To prevent this complication it is urged in the article on castration that a *large opening*, say three or four inches long, be made in the scrotum at the time of castrating, and it is further urged that it be kept open by inserting the hand up into the wound once or twice a day for several days after the operation. However, should an abscess form it is to be opened by a free incision the whole of its length, thoroughly cleansed with water, (warm or cold), then take cotton batting enough to fill the cavity, tie a string around it and saturate it with the following:

Corrosive Sublimate........................ ½ dram
Water.................................... ¼ teacupful

Put the sublimate into a teacup and pour on the water

boiling hot, then wet the cotton in the solution and pack it into the wound. Allow it to remain for 24 hours, then remove it by the string, which is left hanging out. Bathe the parts with white lotion, injecting it into the wound once or twice a day until well.

INTERNAL TREATMENT.

Give one tablespoonful of the following three times a day in the feed or on the tongue:

```
Powdered Nitrate of Potash.......... .....10 ounces
Quinine............................ .......... 4 drams
```

Mix. Feed on a laxative, nutritious diet, and give well regulated exercise. Should the bowels become constipated give one-half pint of raw oil once a day until they are loose.

CHAMPIGNON OR SCHIRROUS CORD.

An enlargement at the end of the cord occurring soon after castration. At times it developes itself quickly; again its growth is very slow. *Causes* are obscure. Cast the animal and dissect out the enlargement and remove it either with the ecraseur or by the clamp. Treatment is then the same as that given for abscess.

GANGRENE.

This complication usually presents itself from the fifth to tenth day.

SYMPTOMS.

Extensive swelling, the parts are cold, and by inserting your hand into the wound you will find it cold to the touch, a loss of sensibility, the discharge is dark and has a disagreeable odor. There is usually an intense thirst, quickened pulse, the visible mucous membrane (lining of the nose) is of a lead color, and the breath offensive. The progress of this disease is so rapid, and the chances of recovery so limited, that treatment must be prompt and energetic.

TREATMENT.

Cast the animal and cut out all the diseased tissue (fl·sh), sear the wound over with an iron at white heat, then sift finely powdered corrosive sublimate into the wound; be su·e that it comes in contact with all parts of the wound; ro'l a ball of cotton batt'ng in the powder, press it into the wound, take a stitch acr·ss the wound to hold the batting in and let him up. Now bathe the swo.len parts with the following:

Aqua Ammonia	4	ounces
Spirits Turpentine	4	"
Alcohol	4	"

Mix and bathe freely morning and evening until the the parts are blistered. Cut the stitches and remove the cotton at the expiration of 24 hours; wait 24 hours longer, and then with a syringe inject the following:

Corrosive Sublimate	½ dram
Water	1 pint

Inject this once a day until light colored pus or matter flows freely, which is a sign of recovery. So long as the matter is of a dark or bloody color having an offensive smell, you have a critical case.

INTERNAL TREATMENT.

Give the following drench at once:

Raw Linseed Oil	½ pint
Spirits Turpentine	1 ounce

Continue this drench once a day until the swelling begins to subside.   You will also prepare at once

Powdered Nitrate of Potash................ 8 ounces
Quinine.................................... 3 drams

Mix and give one tablespoonful three times a day in the feed or on the tongue.   Feed scalded bran mash if he will eat it, with all the pure, fresh water he will drink.   Allow him to pick grass in fair weather, and a loose box stall in stormy weather.   When convalescent (begins to recover) you will give instead of the quinine

Iodide of Potash......................... 4 ounces
Water.................................... 1 pint

Mix.   Dose, tablespoonful three times a day.

MAGGOTS (FLY BLOWN).

Should you find worms commonly called maggots in the wound you are to enlarge it and dig them out with a stick, or what is still better, your fingers, then saturate the wound with spirits of turpentine, aqua ammonia and alcohol, equal parts; mix.   Or you may use

Creosote................................. 1 ounce
Alcohol.................................. 4 ounces

Mix.   Either of these remedies applied once, or possibly twice, will destroy maggots.   Then dress once a day for a few days with compound tincture of myrrh or pine tar.

VETERINARY DENTISTRY,

CHAPTER V.

THE TEETH.

The teeth are objects implanted in and projecting from the maxilliary alveola. They are characterized by the hardness and density of their specific tissue. Teeth vary with the class of animal in number, size, form, structure, position and attachment.

The following table shows the number of teeth in the different domesticated animals:

Animals.	Incisors.	Canine.	Molars.	Total.
Horse	12	4	24	40
Ox	6	2	24	32
Sheep	6	2	24	32
Dog	12	4	26	42
Pig	12	4	28	44
Cat	12	4	14	30

The ox and sheep (ruminants) have neither incisors or canine teeth in the upper jaw, they being supplemented by a hard, cartilaginous pad covered by the mucous membrane of the hard palate (the roof of the mouth) forming a point for the lower incisors to press against while cropping grass, *Dentine* constitutes the greater part of the tooth. *Enamel* is

distinguished by its whiteness, and is the hardest animal tex-
ture. *Crusta petrosa*, or *cementum* completely covers the
imbedded portion of the tooth, and is the softest part, closely
resembling bone. *Teeth* may be simple or compound, simple
as in the dog, where the entire exposed surface is covered with
enamel. Compound as in the horse, where the various tissues
are in wear. Teeth are arranged alongside of each other so
as to form the dental arch. Teeth are of three kinds: The
incisors, or cutting teeth, often called nippers; the *canine* teeth,
or tushes, are situated between the incisors and *molars*, or
grinders.

The horse, like many other animals and man, has two
sets of teeth—the temporary, or milk teeth, and the permanent,
or horse teeth. There are 24 milk, and 40 permanent teeth.
The mare usually has but 36 teeth.

The incisors, or front teeth, in the horse are 12 in number
six in each jaw, the upper ones are the longer, meeting the
lower. In rare cases they overlap the lower and then are
called " Parrot mouth." The younger the teeth the greater
their width, which gradually narrows with age.

AGE AS INDICATED BY THE TEETH.

The teeth undergo a change each year from birth up to
the sixth year. After this the age can only be approximately
determined by the wear in altering the shape of the teeth, circumstances often modifying the wear of them. Hence, after
six years old, an approximately correct opinion can only be
formed by those who have given this subject much time and
thought. The colt is born with two, sometimes three, temporary molars in each jaw. At about 12 months old another
molar, a permanent tooth, appears, and before the expiration

of the second year a fifth molar, also a permanent tooth, shows
itself.   At about two and a half years old the two first temporary molars are replaced by permanent teeth, and between
three and four the remaining temporary molars are replaced
by permanent teeth.   About this time the last, or sixth, permanent molar begins to appear.   The molars are seldom referred
to as an index of the age on account of their location   Never-
theless it is useful to be acquainted with their changes.   After
four years old the molars are not taken into consideration in
determining the age.

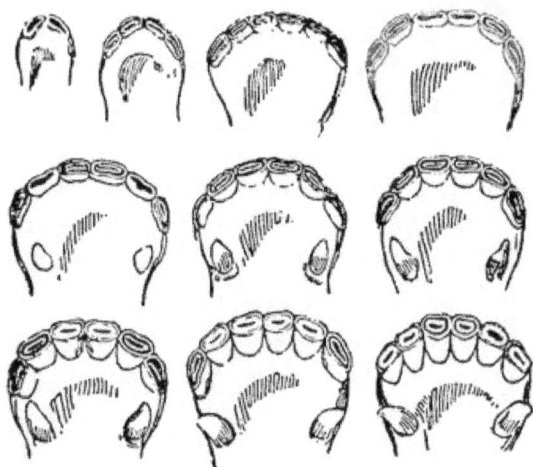

[THE TEETH.]

THE MILK TEETH.

The colt is born with his front teeth covered with the
gums, and at various periods during the first eight or ten
months the different temporary incisors appear, which is very
beautifully illustrated in the following verses.

THE HORSE'S AGE.

To tell the age of any horse
Inspect the lower jaw, of course,
The six front teeth the tale will tell
And every doubt and fear dispel.

Two middle "nippers" you behold
Before the colt is two weeks old ;
Before eight weeks two more will come;
Eight months the corners cut the gum.

The outside grooves will disappear
From middle two in just one year,
In two years from the second pair;
In three the corners, too, are bare.

At two the middle nippers drop;
At three the second pair con't stop :
When four years old the third pair goes;
At five a full new set he shows.

The deep, dark spots pass from view,
At six years, from the middle two;
The second pair at seven goes;
At eight the spots each corner lose.

From middle nippers, upper jaw,
At nine the black spots will withdraw;
The second pair at ten are white;
Eleven finds the corners light.

As time goes on the horsemen know,
The oval teeth three-sided grow;
They longer get, project before.
'Till 20, when age is known no more.

GENERAL REMARKS ON TEETH.

Having briefly described the teeth and their changes, I will now proceed to speak of a few of the many ails incident to the horse's mouth and head in consequence of diseased teeth and the irregularity of their wear.

There is no part of a horse that requires more care and attention than the mouth, and the cow is not to be excluded. The horse being mute cannot tell us of his aches and pains, yet if we give him the attention due any faithful servant we are sure to note his complaining by the various attitudes assumed while eating, drinking and driving. Many horses die annually from indigestion, the primary cause being faulty teeth. This statement may seem strange to many who have never chanced to examine the horse's mouth. The wear of the teeth is such that the inside of the lower and the outside of the upper teeth are left sharp. Thus it will be seen that in the first instance the tongue is wounded during mastication,

and in the latter the cheek is wounded at times to a considerable extent. When the mouth becomes thus mutilated the
food is taken into the stomach in a very crude state, the
results of non-mastication, in consequence of which the stomach
is overtaxed. Following this we have indigestion, colic,
flatus, etc.

ULCERATED TEETH.

The first symptoms usually are a swelling which is attributed to a kick or knock on the jaw or face. In the course
of a week or two this swelling suppurates and breaks discharging matter or pus which has a fetid, offensive smell. Sometimes the ulcer of an upper tooth will open into the nasal
cavity and discharge from the dostril. This condition has
often been mistaken for nasal gleet and glanders.

While I do not suppose for a moment that a stock raiser
or farmer is going to attempt the extraction of these teeth,
yet we make mention of these facts that they may be on the
lookout.

SYMPTOMS OF BAD TEETH.

At times a horse will carry his head to one side while
being driven "*Side Line.*" Again you will see an animal
while eating grain, extend the head, turning it over sidewise
while in the act of chewing, or you may see a horse take hay
into his mouth, chew it a while and then spit it out ; other
times the coat will be staring and unthrifty, the results of
indigestion brought about by bad teeth. A discharge from
the nose of a fetid, offensive smell, or a discharge of stinking
matter from a sore on the under jaw are all signs of bad teeth,
and should you see any one of the foregoing symptoms I

would advise you to consult a veterinary dentist at once, thus relieving the suffering of a poor dumb animal that serves you more faithfully than your domestics.

WOLF TEETH.

[THE WOLF TOOTH FORCEPS.]

These are supernumerary teeth and are usually situated just in front and close to the first upper molars. However, these teeth may be found in any part of the body. I once found one in front of and close to the horse's ears, and Uncle John Steiner. of Bluffton, Ohio, once found one i n the spermatic cord of a horse near the testicle.

There has been much said and writen about the wolf tooth and its influence over the eye. My opinion is that if the tooth is an extra long one, it will effect the eye, but if a short one, as is usually the case, no harm can come from it. However, as they do no good, I should advise their removal by extraction and not by breaking them off. The above cut shows the forceps used for this purpose.

DRESSING THE TEETH.

[THE FLOAT.]

When you have a horse with sharp projections on the

outside of the upper and inside of the lower teeth, you are to remove them with the *Float* by carefully filing them off. Do not file the face or articulating surface of the teeth, but the edges alone. There are men traveling the country as veterinary dentists, who leave the mouth in a much worse condition than they find it by overdoing ; that is to say by rasping the grinding surface too much. When your animal has once been subjected to the maltreatment of these vile pretenders, nothing but time and lots of it can restore the fine grinding surface of the tooth to its natural condition.

TREPHINING.

[THE TREPHINE.]

There are some diseases of the head and teeth that cannot be treated without trephining (cutting a hole through the bones of the head) with the above instrument. This operation should never be undertaken except by a duly qualified veterinary surgeon. However, there are some difficulties that can be remedied by the use of the float alone.

LAMPAS.

A mild inflammation of the bars of the mouth causing them to swell to an extent which may interfere with mastication. Take a sharp knife and make a few cuts across the

the first two bars say five or six marks sufficiently deep to cause bleeding and bathe with alum water. Do not burn them nor allow anyone to burn them for you, as it is not only a useless, but a barbarous practice.

THE CONKEY INCISOR CUTTING FORCEPS.

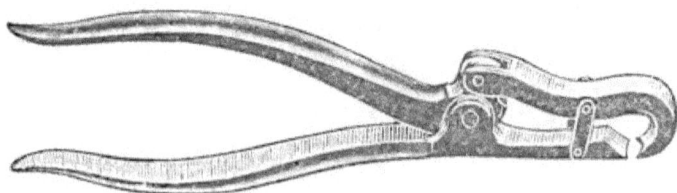

[PATENT APPLIED FOR.]

With this instrument the incisors or nippers of a horse can be cut off in five minutes. They are easily worked with one hand, and do not split or mutilate the teeth, but leave them nice and smooth. For particulars, address,

L. L. CONKEY, V. S.

Office No. 6 Canal St.,

Grand Rapids, Mich.

REFUSING TO EAT WITHOUT DISEASE.

If a horse refuses to eat without any apparent signs of disease, you had better examine his mouth well back to the roots of the tongue for a corncob or other hard substance, which may be lodged there. On several occasions I have found a piece of corncob lodged between the back teeth crosswise of the mouth. These animals have been treated for all kinds of diseases by local talent without the slightest idea of what was wrong.

THE EYE AND EAR,

CHAPTER VI.

INFLAMMATION OF THE EYE.

If the inflammation be due to external injury as a blow, the stroke of the whip, the bite of an insect, the mark will usually show itself. The cause of an injury to the eye at times is a matter of importance, as a man's character may be at stake, the owner blaming his employe for striking the animal and causing the disease. In such cases the veterinarian is appealed to by both parties. The symptoms are usually quite plain, there is swelling and redness with tears flowing from the eyes, and often we find an abrasion (wound) of the external covering of the eye which may look at first quite serious, but if the proper treatment is adopted the results are surprising.

TREATMENT.

Remove the cause of irritation, then bathe, in mild cases, with warm water, but the more severe cases will require bleeding from the *angular vein* below the eye; make a large open-

NOTE.—You must not confound this disease with the conditions somtimes met with in influenza: the purgative in that case might kill.

ing and allow it to bleed freely for sometime, and in most cases you can allow it to stop of its own accord. After the bleeding stops hang a piece of light cloth over the eye fastening it into the halter with holes for the ears, then wet an extra piece of cloth in the following and lay it over the eye under the cloth already affixed:

Atropine Sulphate.........................10 grains
Water.................................... 1 pint

Mix. Apply three or four times a day. Give a purgative ball at once followed by a few ten drop doses of aconite, then a few tablespoon doses of nitrate of potash. A dark stall is always to be preferred for horses suffering with diseased eyes.

PERIODICAL OPTHALMIA.

MOON BLINDNESS, CONSTITUTIONAL OPTHALMIA.

This disease is of a transitory character at first, but sooner or later comes to stay. This disease is not so common as it used to be in the days of log stables without windows. Some attribute this decline to the veterinarians, but I think it is due to better stabling.

SYMPTOMS.

It usually comes suddenly and in the night. The eyes appear to be weak, the lid is drawn down a little, while the eyeball is drawn back in the socket giving it the appearance of being smaller than its mate. The inflammation is apt to move from one eye to the other, then in a few days disappear altogether, but often leaves the eyelid in a peculiar wrinkled condition. The eye may remain free from active inflammation

for a time, when the animal is again attacked, the eye is
closed, weeping and apparently suffering intense pain. Again
the disease may disappear for weeks or monthss and
then return, this going and coming may be of short durations,
or marked with long intervals, but the final results are the
formation of a cataract in one eye and then in the other until
the animal is totally blind. Prof. Smith, of the Toronto
Veterinary College, says: "Never breed a mare to a horse
suffering from periodical opthalmia."

TREATMENT.

Is only palative and consists in giving a *purgative ball*
and bran mashes followed by frequent doses of *epsom salt*
(quarter pound), bathing the eye often with

Zinc Sulphate	2 drams
Fluid Extract Belladonna	2 "
Loaf Sugar	2 "
Water	1 pint

Mix and bathe frequently in the same manner as recommended in *inflammation of the eye*. Keep the animal in a
well ventilated dark stall ; avoid overwork and fatigue.

DISEASED HAW.

The *membrane nictitans*, commonly called the *haw*, which
is the horse's pocket handkerchief used in removing foreign
bodies from the eye, sometimes becomes inflammed caused
by an irritation of some kind ; the haw is red and swollen.
This condition has often been termed "*the hooks*," and the
haw used to be cut out by the ignorant pretender. Whenever
a horse is found suffering with an inflammation of the haw

use warm water applied with cloths in the same manner as described in inflammation of the eye after fomenting for a day or two, use the following for a week :

Zinc Sulphate..........................	1 dram
Sugar Lead.............................	½ "
Loaf Sugar.............................	2 "
Water.................................	1 quart

Mix and bathe frequently.

WARTS IN THE EYE.

These are sometimes very troublesome and may cause blindness. They are of two varieties in the horse and a third of a malignant form in the cow

SYMPTOMS.

The most common form met with in the horse is the red, or blood wart, which makes its appearance in the lower corner of the eye, and may grow very rapidly, soon covering the whole eye. The second variety is of a black color ; this also grows in the lower corner, and like the red wart, it is attached to the haw. This is called a melanotic tumor, and is common to white or grey horses.

TREATMENT.

First cast the animal in the most convenient way, then take the curved needle armed with a cotton cord, and pass it through the wart; the string is to draw the wart out and hold it while you are cutting it loose with the shears. In most cases you will have to remove the whole *membrane nictitans*, and you are pretty sure of a complete recovery. After the removal you are to bathe the eye frequently with the following:

Zinc Sulphate. .... 1 dram
Tincture Belladonna....................... 1 ounce
Rain Water................................ 1 pint

This is a delicate operation and should be entrusted only to a skilled operator.

EAR—WARTS.

The most common ail to which the ear is subject is warts. These are of various kinds, and their removal is easily accomplished with the knife, and usually without hampering the animal more than applying the twist to the nose, when you will take hold of the wart with your hand or pass a needle through it drawing a string through for the purpose of drawing the wart away from the skin. Now take a sharp knife and cut the wart off; be sure to remove all of its roots from the skin. There is not much bleeding following the removal of a wart from the ear. However, you might sear the wound over with a hot iron, and the *bulb iron* answers well for this purpose.

DEAFNESS.

The causes of deafness in the horse are obscure. In the army it comes from the noise produced by the guns, but I have seen a number of deaf horses that perhaps never heard the report of a gun. The symptoms are a general looseness of the ears, the animal may occasionally lift his ears in a natural position, but quickly drops them again, when they hang outstretched from the head on either side and a little forward. It is barely possible that a horse that is not deaf might carry his ears in this manner, but if I saw a horse with his ears in an apparently lifeless condition I should examine him for deafness before passing an opinion as to soundness.

OBSTETRICS AND PARTURITION.

CHAPTER VII.

GESTATION.

The period that the mother carries her young varies considerably in different animals, and even in the same species there are variations, though not very great yet ot importance. The female elephant carries her young *two years*, and the female rabbit 28 *days*. The average period of the MARE has been given at 11 *months*; the longest period 400 and the shortest period 300 days. However, my own observations have placed the average period at 11 *months and 15 days*, and the foals that are carried longer are usually weak, crooked legged and deformed. Mares that are well fed carry their young longer than those in poor condition, and a mare bred to a thoroughbred stallion goes longer than a mare bred to a common bred horse.

THE COW.

It is commonly stated that the cow is pregnant the same length of time as a woman, which is true of the average, although the cow varies to some extent. In the "*American Journal of Medical Science* for 1845, in reporting 62 cows, gave the longest period as 336 days, and the shortest as 213 days. The average for male calves being 288 days, and females 282 days." The farming community in general have an idea that the male calf is carried much longer than the female, but I do not think that sex plays any part in the period of gestation in the cow.

SHEEP AND GOAT.

The sheep and goat carry their young about *five months,* the average period of the ewe being 149 days the longest period being given to the ewe lambs. With regard to breeds, our best authors say that the "Merino average 150 d, 3 hrs, and the Southdowns only 144 d., 2 hrs, or about six days less.

The goat goes a few days longer than the sheep.

THE PIG.

The pig is usually pregnant *four months,* although some authors gives the period at *three months, three weeks and three days.* However, the average is 119 days.

THE BITCH.

The bitch carries her young from 58 to 65 days.

THE CAT.

The cat is pregnant from 50 to 60 days, the average being 55 days or eight weeks.

PLURIPAROUS—TWINS.

Of all domesticated animals the mare is the one that less frequently brings forth twins, and these are dead, or soon die, although I have seen a few pairs of twin colts that lived and did well.

The following report may serve to interest the public: "The most numerous instances of twins in the *mare* are, however, to be attributed to two successive fecundations, of which Saint-Cyr has collected eight examples. In all of these, strange to say, the mares had been put to a stallion of the equine and asinine species in succession, and brought forth each a foal and a mule."

The cow often brings forth two or three calves, and Fleming reports a small cow of the black polled breed that brought the following number of calves. In six years she had 25 calves, as follows: 1, 3, 4, 2, 3, 6, 2, 4, seven of which died. She calved eight times in six years, and raised 18 calves, an average of three calves a year. Kleinschmeid reports having found fifteen *little* calves in the uterus of a cow.

TWIN LAMBS.

This is very common and in a good flock of sheep there should be as many lambs as ewes, the doubles making up for the loss. "In May, 1876, a farmer had some ewes fattening and a neighbors ram got with them and 13 ewes got with lamb and produced no less than 31 lambs, all born alive, as follows: One, 1; eight, twins; three, thriplets; one, 5—31." It may be remarked that this extraordinary number is seldom seen in high-bred sheep. It appears to pertain to the common breeds. Not only this but certain years are more remarkable than others for double, treble, and quadruple births in sheep.

FREE MARTINS.

A curious fact in connection with twin calves is that if

they are of both sex the female is usually unproductive, and for this reason they have been styled "Free martins" in English, "Queenen" in Holland, "Zwitter," or "Zwilling" in German, "Loures" in French. The old Roman agriculturalists called these animals "Taurae." There has also been some doubts about the male being productive. One case is reported where a twin bull served one hundred cows, none of which proved with calve. Many similar reports are recorded, but this will suffice to give you an idea in a general way.

THE CARE OF PREGNANT ANIMALS.

The *mare* should not be worked too severely or fatigued particulary as the term advances, and on the other hand absolute rest is dangerous. *Exercise* is absolutely necessary and most, or all, cases of difficult parturition occur among those deprived of exercise. Mares in foal will perform ordinary work with benefit. Do not allow a mare in foal to stand in her stall for several days, or weeks, and then begin work by hauling heavy loads, and in no case allow her to be jerked suddenly, shocked or frightened. The food should be good and in abundance. Toward the end of the term scalded bran mashes, oil cake meal, etc., are to be given quite freely, and an occasional dose of raw oil (one-half pint) will be found beneficial along about foaling time. The cow is to be fed on a laxative diet, roots, scalded bran mash, etc. If the cow is fat you had best give her a good purgative two or three weeks before calving (See Apoplexy;) this may be the means of saving your cow's life, while a laxative, nutritious diet, something to oil up the parts and assist nature in the great work she has to perform may be the means of saving the life of a valuable colt or calf. A penny's worth of preventative is cheaper than a pound of cure, and this is the time to apply it.

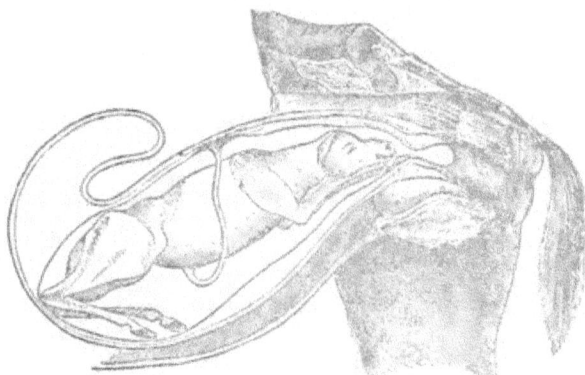

THE NATURAL PRESENTATION AT FULL TERM.

The above engraving shows the position assumed by the fœtus of the different domesticated animals at full term and any deviation from this position is unnatural and will require the assistance of man to complete delivery. However, the variations are numerous, some of them are mere trifles, while others are of a very grave nature. One leg may be crossed over the neck, one or both front legs may be completely retained or bent at the knee, again the head may be bent upwards or downwards, to the right or to the left and backwards the length of the neck. This is somewhat difficult to overcome in the mare, but comparatively easy in the cow on account of the difference in the length of neck.

DR. CONKEY'S OBSTETRICAL SET.

This set of obstetrical instruments are all made to fit the one handle, and when joined together are 30 inches long.

These instruments are the results of the last 10 years
practice having been improved upon from time to time, the
last improvement, the shield, was invented during the present
month, January, 1890, and I feel that I can now say to the
public that I have the only thoroughly practical set of instruments in the market. They are made of the best steel and

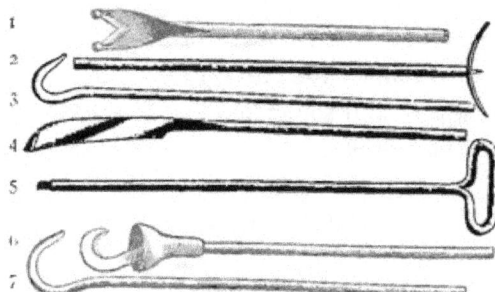

[PATENT APPLIED FOR.]

Figure 1, Guarded Chisel Knife; 2, Repulser with removable spike; 3, Sharp
Hook; 4, Saw; 5, The Handle; 6, Sharp Knife with Funnel Shield; 7, Blunt
Hook.

guaranteed in every particular. The knife is so constructed
that it is almost impossible to cut a mother. Price per set of
seven pieces, nickle plated, $15, net. Address Dr. L. L.
Conkey, Grand Rapids, Mich. I always carry two
sharp and one blunt short ring hooks. These are nickle
plated and will be sent on receipt of $1 each.

NATURAL POSITION WITH THE HEAD BENT TO
THE LEFT.

This is perhaps the most common grave deviation met
with, and is not so hard to overcome in the cow, but on
account of the length of the colt's neck it often gives the operator a hard task, not infrequently resulting in the death of

both mother and foal. However, the foal is usually dead
before aid is called, on account of the connection between the
foal and the mother being severed with the first labor pain,
and it is acknowledged that the foal cannot live past the sixth
hour thereafter, but often dies in less time. The connection
between the mother and calf differs and is never severed until
the calf leaves the mother altogether; for this reason the cow
may be in labor for several days and the calf be born alive.
Again, difficult labor of the *mare*, in which pulling and working has been severe and somewhat contracted, the results are
often unfavorable, while in the *cow* the reverse is the case.

EMBRYOTOMY.

THE ACT OF TAKING THE YOUNG BY CUTTING IT AWAY.

I will endeavor to describe the mode of taking a *colt*
away when the head is retained and bent to the left, as shown
in the cow (See Engraving next page), which is as follows:
Stand the mare in a stall, put a halter on, run the tie strap
through the usual place in the manger, do not tie it, but give
it to an assistant, who will stand near the shoulder and endeavor to keep the mare quiet. A second man will stand at
the hips to keep the tail out of the way and hand you the
instruments. Now take off your coat, vest, and shirt, put
on an old vest, soap or grease your arms and you are ready.
Fasten a cord or small rope to each pastern of the colt, tie
a third cord to one of the short blunt hooks and endeavor
to hook it into the colt's mouth or eye. Should you
succeed you will now push the legs and body back as
far as possible holding it there with the repulser (No. 2),
fixed by the screw to No. 5, and held by an assist-

ant; now pull the head out as far as possible, when you can reach the mouth you will fix a loop over the nose and through the mouth and continue your pushing and pulling until the head is straight when it will come away with

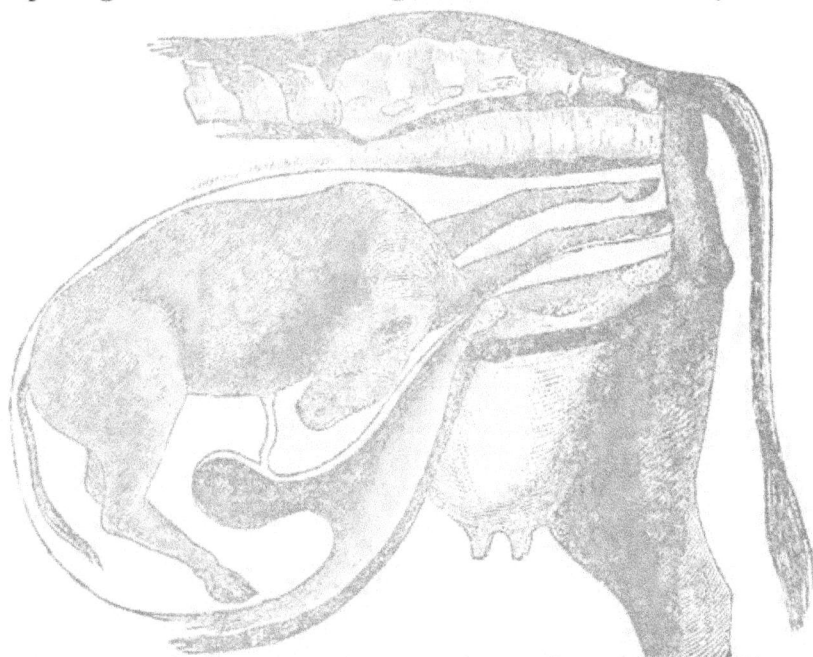

little or no help. This, however, is much easier said than done, and if you fail in straightening the head you will adopt the following: Have the legs drawn well out, fix the knife (No. 6) to the handle (No. 5), take the knife in your left hand and the handle in the right, carefully introduce the knife, keeping it in your hand until you reach the left shoulder of the colt, then hook it into the shoulder of the colt and carefully withdraw it cutting the skin open to the foot; now lay the knife away and with your hand loosen the skin from the leg; do not cut the skin off at the foot until you have loosened it all around the leg to the body, beyond the elbow; now cut the skin off below the knee, have your assistants pull on the left leg which will tighten the muscles, now fix the chisel (No. 1) to the handle, take it in your left hand and introduce it for the purpose of cutting the muscles between the leg and breast bone of the colt. When this is done one good, smart pull will bring the leg away. Now renew your attempt to get the head, and if you again fail which is apt to be the case, you will cut the head off. First use the knife to cut the upper muscles to the bone, then fix the saw (No. 4) to the handle and pass it in carefully guiding it with your left hand; when you have the saw in the right place take hold of the neck with the left hand and do the sawing with the right or open air hand; saw the neck bone off, then take the knife and cut the remaining muscles of the neck and bring the head away. Now stop and oil the parts well with raw linseed oil or melted lard using the injection funnel. After you

[THE INJECTION FUNNEL.]

have oiled the parts well gently pull on the leg of the colt, keeping one hand over the stump of bone on the neck to prevent it from wounding the mother. When you have the colt away put one teaspoonful of permanginate of potash into a pailful of blood warm water and wash out the womb again using the injection funnel. Now give about four ounces of whisky, make a dry bed, close the doors and leave her alone for five or six hours, when she should have a warm bran mash and a pail of chilled water. If the labor has not been too severe she will require nothing more than ordinary care, but you must watch her closely for a few days as certain complications may follow, though the operation may have been performed without difficulty.

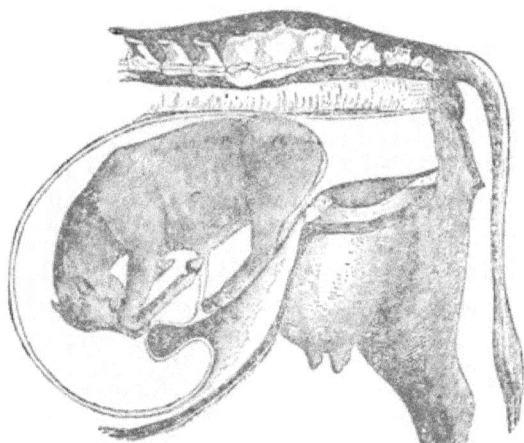

CROUP, OR BREACH PRESENTATION.

While this is not so difficult in the cow, it is a herculean task in the mare. The first thing to be attempted is to pass a rope between the hind legs and bring it around the outside of the leg into the open air (keep the rope thoroughly greased with lard) and forming a slip-noose tightening it around the leg as you work it down toward the foot; be sure to work it below the hock, and the fetlock is better but is difficult to get it lower than the hock at first. After you have one hock corded, cord the other in the same manner. As soon as you have both hocks secured you will station five men as follows : No. 1 at the head, No. 2 at the hip holding the tail out of the way, No. 3 at your right, No. 4 at your left, and No. 5 at your back to help you push. If the pains are excessive give the mare one-half ounce of chloral hydrate in half pint of water. Now as you push the colt forward No. 5 is to brace against you assisting as best he can. As soon as you feel the colt going forward call on No. 3 to pull quick and steady backward and upward; now push again, calling on No. 4 to pull backward and upward. Nos. 3 and 4 should stand on a box, or some object 18 inches high. If you watch and take advantage of the time when the mare does not labor you will soon raise the hock joints into the pelvic cavity and there is nothing to prevent you from doing this when once you get a rope around the legs. After you have the hocks raised slip your rope down below the fetlock and proceed as before. When you have both legs out you will *oil* the parts thoroughly before you commence pulling, and then pull only when the mare labors. Hold what you have until she labors again. Continue this; pulling carefully until you get the colt away.

SECOND MODE.

Should you fail to get a rope around the legs you will adopt the following: Introduce the sharp knife (No. 6) into the rectum of the colt cutting the flesh away until you can remove the bowels, lungs, heart, etc., then attach a short ring hook to the pelvis bone; cut and pull, pull and cut, until you get the pelvis away, next take one leg and then the other. Continue in this manner until you have the entire colt away. It does not take long to tell this, but it requires about two hours' hard labor to accomplish the operation.

GENERAL REMARKS ON TAKING COLTS AND CALVES.

The manual labor in taking colts and calves from their mother depends much upon the different parts presented as

well as the efforts and condition of the mother. The first
thing to consider is the regions of the body presented to the
external opening. The next thought is the relative size and
shape of the part presented compared with the opening.
Again we must not overlook the symptoms afforded by the
labor pains, as they are not always in harmony with nature,
for we not infrequently meet with cases where the *labor pains*,
or contractions, assume the opposite directions of those occurring in healthy labor. This is called

TUMULTOUS LABOR.

Symptoms.—The pains are excessive and frequent, yet no
progress is made, the parts do not seem to be prepared,
the mouth of the womb seems to be in a state of spasmodic
contraction, hard and painful. The pains commence at the
mouth of the womb and pass back, forcing the fœtus toward
the fundus. Tumultous labor is mostly confined to the primi-
paræ.*

TREATMENT.

Give gentle exercise for an hour, then place her in a dark,
quiet stall and give of

Chloral Hydrate........................... ½ ounce
Water...................................... ½ pint

Then wet a sponge with fluid extract of belladonna and
place it in the vagina back against the mouth of the womb,
renewing the medicine every six hours. You will introduce
your fingers into the womb, carefully dilating its mouth as
often as you change the sponge. The calf will usually come
away in 6, 12 or 24 hours after the irregular labor subsides.
This ail is liable to be confounded with

---

* One which has never before brought forth young.

INDURATION OF THE CERVIX.

This term is applied to an altered condition of the mouth of the womb, and means in plain English that the mouth of the womb has lost its elasticity and will not open. It has a leathery feeling, but is not hard, hot and sensitive as in *rigidity*. Induration is only met with in the cow.

TREATMENT.

First endeavor to dilate the opening with your hand, but if you fail you will then have to make three to five cuts through the leathery part of the womb. The cuts are to be made each side of the center, and not in the center under any consideration. After you have taken the calf away, wash out 'the womb, using the injection funnel with very warm water, using one or two teaspoonsful of permanginate of potash in each pail of water. This will usually stop the bleeding. Should there be much bleeding you are to give two to three ounces of tincture of opium in a little water, and you may have to pack the womb full of clothes saturated in vinegar or alum water. However, I have never seen any bleeding after the hot water and potash was used. Give a few doses of quinine and attend the general comfort.

RIGIDNESS, OR SPASMS OF THE WOMB.

The *symptoms* vary somewhat. The only thing that is likely to awaken suspicion is the unusual duration of labor which may extend over two, three or four days, if assistance is not offered. In other instances the animal manifests an unusual amount of excitement. The only other condition with which *rigidity* might be confounded is *induration*, but in rigidity the parts are hot, tense and painful, and without the leathery appearance of induration.

TREATMENT.

Give rectinal injections of

Warm Water............................ 1 gallon
Tincture Belladonna.................... 2 ounces

Then take a good sized sponge and wet it with fluid extract of belladonna, and put it into the vagina as far as you can force it, leave it there six hours, then renew the belladonna and return it. Keep the animal in a dark stall by herself and as soon as the spasms relax the calf will come unassisted.

INFLAMMATION OF THE WOMB.

This is usually the result of, or follows parturition, especially if the labor has been prolonged and difficult ; lying on damp ground, or exposure is liable to cause it.

SYMPTOMS.

It usually occurs the third or fourth day after foaling The back is arched, urine passed frequently in small quantities, sometimes they will lie down, groan and occasionally look anxiously at the side, the pulse 80 or 90 beats per minute, mouth hot, ears cold, or alternately hot and cold, there is usually stamping or paddling of the hind feet and a whisking of the tail.

TREATMENT.

First give one pint of raw linseed oil, then give 10 to 15 drops of fluid extract of aconite root every four hours alternate with nitrate of potash, a level tablespoonful every four hours; this brings the medicine two hours apart. If there is a discharge from the vagina wash it out with warm water, using a little permanginate of potash—enough to color the water red. Repeat this once a day as long as the discharge lasts, using about a pailful at each time. Apply the *hartshorn liniment* to the back, and if the legs are cold bathe them lightly also, then bandage. Keep her very warm, offer chilled water often and feed scalded bran mash and vegetables.

PARTURIENT APOPLEXY.

MILK FEVER, CALVING FEVER, ETC.

This disease may properly be said to affect a certain class of cows only. Fat, easy keeping, heavy milkers are the ones subject to it, and it occurs from the third to the seventh calving. The act of parturation (calving) seems to be the cause. Most writers say that apoplexy makes its appearance from one to three days after parturition. Flemming says that it may attack a cow as early as twelve or twenty hours after parturition, but is most frequent on the second or third day.

Out of twelve cases treated by me, five were attacked between the twelfth and twentieth hours after parturition. Seven were attacked between the twentieth and the thirtieth hours. Those attacked early died, while the latter seven recovered. This leads to the following conclusion: That the earlier the attack, the more fatal are the results, and the later the attack, the more favorable the results.

SYMPTOMS.

The first thing generally noticed will be a paddling, staggering gait. Some cases have been reported to me as " wabbling behind." These symptoms are speedily succeeded by loss motor power, the cow lying or falling down ; she may make an effort to get up, but usually fails. Sometimes they are very restless, throwing themselves violently, but usually lay quiet, either outstretched with the head on the ground, or in a natural position with the head thrown around against the side.

All these changes may take place in a few hours. An animal which was left in apparent good health a short time before, is found lying, cannot get up and is in a comatose (sleepy) condition. You can put your finger against the eyeball without the animal evincing any pain. The breathing, at first accelerated, soon becomes slow, deep and stentorious. Bowels are costive, and urine scanty.

TREATMENT.

First place the animal in a natural position by boulstering her up with sacks of straw, keep the head above the body by means of ropes. Do not allow her to lay stretched out on her side, but keep her lying on her sternum (breast bone) as a cow usually lays while ruminating (chewing the cud). Then take from four to six quarts of blood from the jugular vein, make a large opening that the blood may flow freely. After pinning up the wound you will give a purgative, composed of

Aloes	2 ounces,
Gention	1 "
Croton Oil	½ dram.

Add molasses enough to make this into a stiff mass, divide into three balls, give one every five minutes until you have given all three. Oil your hand and push the balls as far over the tongue as possible, where they will slowly gavitate into the rumen (stomach).   You will now take of

Aqua Amonia............................. 4 ounces,
Spirits of Turpentine..................... 4   "

Mix, shake and bathe the back all along on both sides of the spine for a space of twelve or fourteen inches wide including the neck, also bathe the legs freely, giving plenty of hand rubbing, then blanket, this promotes circulation.   You will now oil your hand and pass it into the rectum and remove all the excrements that you can reach, then give warm water injections freely and often.   No harm can possibly come from injections, even in large quantities.   Offer her water to drink often, say once an hour.   After you have the cow nicely under treatment, you will give

Fluid Extract of Noxvomica................ 1 dram,
Whisky......... .... 4 ounces.

In a half pint of warm water, every four to six hours, until she shows signs of improvement.   I have always made it a practice to draw the milk hourly and consider it a good symptom to get milk, be the quantity ever so small.

PREVENTATIVE TREATMENT.

About four weeks before calving give the cow a good purgative.   For this purpose take one pound of epsom salt, and boiling water enough to dissolve it, add one pint of molasses and pour it all down her at once.   Then repeat the dose a week before calving.   Your only safety from this dreaded malady is through preventative treatment.

DISOWNING ITS YOUNG.

It sometimes happens that the mother will disown its young and you have a great deal of trouble to get her to allow it to suck. I have frequently overcome this by rubbing the young over with the placental membrane (after birth), which causes the mother to lick it and take to it more kindly. In this way you can rub over a strange lamb with the afterbirth of a ewe and she will generally own it and allow it to suck.

ŒDEMA—"SWELLING."

This has often been dubbed *farcy*. In fact, nearly all swellings are styled *water farcy* by the horse doctors who are too lazy to read. Oedema is characterized by a swelling of the hind legs and belly in front of the udder. This swelling may be confined to the legs only, or it may be confined to the belly, sometimes slight, at others very great. Well bred animals are not so susceptible to it as the coarser bred. It is caused by the heft of the colt pressing on the abdominal bloodvessels preventing a free return of the blood, thus impairing circulation, causing an extravasation of the blood into the surrounding tissue. The swelling will usually disappear with exercise. There is nothing serious about this œdema of pregnant animals as it can usually be counteracted by scalded mashes and an occasional dose of oil with well regulated exercise, and will disappear in a day or two after foaling.

RETENTION OF THE AFTERBIRTH.

This accident in the mare must be attended to at once. The removal if effected by taking hold of the afterbirth with one hand, pulling it enough to give it a tension, then carefully work the fingers of the other hand between the afterbirth and the womb separating them carefully; something of the same manner that you would remove the pelt of a sheep. After you have the afterbirth removed you are to wash out the uterus with *red lotion*. Retention of the afterbirth is much more common in the cow than the mare. It is not at all uncommon to see cows, eight or ten days after calving, that have not got rid of their afterbirth, and yet they are lively, chew their cud, and give a good quantity of milk.

TREATMENT.

There is a great diversity of opinion on this subject, some claiming that it should be removed by manual labor. This may be true in some cases, but for me, I prefer the following treatment, which has, I think, given universal satisfaction.

UTERINE TINCTURE.

Powdered Savin.........................	2 ounces
" Cumin...........................	1 "
Treacle..	2 "
Essence of Rue...........................	1 "
" " Savin	1 "
Alcohol...................................	20 "

Mix and bottle for use. Of this tincture give one ounce in a half pint of raw linseed oil morning and evening until the afterbirth comes away. This remedy has always done the work for me, and I consider it second to none.

IMPRISONED AFTERBIRTH.

It sometimes happens that the afterbirth gets broken off, and I have known a few smart *cow doctors* who would pull the afterbirth out as far as they could handily and cut it off, when the remainder would drop back into the womb and the neck or mouth of the womb would close over it imprisoning the undetached portion, which would decay and come away in stinking, mattery fluid.

SYMPTOMS.

The cow runs down in flesh, falls off on her milk, back arched, with frequent straining as if to urinate when a quantity or purulent matter will escape.

TREATMENT.

Wash out the womb once or twice a day with the *red lotion*, using the injection funnel, and give internally the following:

Ergot of Rye........................... 4 drams
Raw Linseed Oil......... ....16 ounces

Mix and give all at once followed by

Powdered Nitrate of Potash.............16 ounces
Quinine................................ 1   "

Mix and give one tablespoonful three times a day on the tongue or in the feed.

ABORTION.

An abortion is the expulsion of the young before it attains sufficient development to live external to its parent

Saint-Cyr says that it may be acknowledged that abortion has taken place when the young is expelled in the mare before the three hundredth day, in the cow before the two hundredth, in the sheep before the hundred and fortieth, and in the pig before the one hundredth day.

The cow and mare are the ones that most frequently abort. *Causes* of abortion are numerous. Irregular seasons, cold, when suddenly applied to the skin, may produce it; cold nights and warm days has an influence, food of a bad quality, the smut of wheat, rye, and corn may produce it; allowing animals to fill themselves with cold water when exceedingly thirsty, is another cause. Rue, savin, ergot, opium and digitalis are to be administered with care. Excessive exertion, especially if the exertion is sudden and severe, or even moderate, if coming after a long period of rest; fright or excitement is also liable. Some say that the smell of blood will produce abortion, but I do not think the sight or smell of blood, without fear, would have any bad effect although the nervous excitement caused by the sight or smell of blood may, and often does, produce abortions. Excitement, fear, sudden surprise, anger, heavy thunder and certain odors are all fruitful causes of abortions.

Should abortion be produced during the first half of pregnancy, the effect on the mother is scarcely noticeable, but if it occurs during the last half, it is more serious, and the mother should be looked after for several days, as there are many complications liable to follow, such as inflammation of the womb, prolapsus of the womb, founder, etc. The udder must be looked to, as the milk may start and the udder become inflamed or caked, (see mammitis).

MAMMITIS.—"CAKED BAG."

This is a common disease and may attack any and all domestic animals, however the cow, mare, ewe and goat are the ones most often effected. First we have congestion, followed by inflammation, sometimes the whole udder, at others one-quarter only is effected. The quantity of milk is diminished, and often has a thick creamy appearance. This disease has often been mistaken for inflammation of the udder, from which it may be distinguished by the lack of constitutional disturbance, as well as the character of the milk. The udder is not so sore and tender to the touch as in inflammation.

Mammitis requires no special care if the animal eats well, you are to draw the milk every two hours or allow the calf to suck. Smearing the udder twice a day with the following: Powdered gum camphor, two drams; hogs lard, two-thirds teacup; melted together, when cool it is ready for use.

INFLAMMATION OF THE UDDER.

Though not so common as mammitis this is much more severe, and occasionally destroys life in a few days.

The first *symptoms* do not differ much from mammitis, but in a day or so the animal becomes dull, ceases to ruminate, the bag is hard and painful to the touch, the quantity of milk diminished, thick, lumpy, and perhaps of a pink or bloody color. The prognosis is usually unfavorable.

TREATMENT.

The general treatment must be that of inflammation. Give a laxative.

Epsom Salt	8 ounces
Castor Oil	8 "
Warm Water	1 pint

To be given at once followed by 15 drop doses of fluid extract of aconite alternate with tablespoon doses of nitrate of potash, every four hours; this brings the medicine two hours apart.

External Treatment.—Take a piece of cotton cloth one

yard square, double it together forming a triangle, called a *support*, tie a stout cloth string to the two long ends of the support, which are to be tied around over the back in front of the udder, four holes are made for the teats and a third long string with its center tied to the remainining back point of the support; this forms two strings which are to be brought back between the hind legs and crossed over the cow's croup and tied to the strings already in place, and tied tight enough to support the udder. Now wet sponges or cotton cloths in hot water and place them inside of the support, keeping the udder warm for 48 hours, then remove and smear the udder over with belladonna ointment, or fluid extract of belladonna and replace the support, packing it with dry cotton batting. Draw the milk every two hours until relieved. You may have to repeat the laxative in a few days.

COLT AND CALF DISEASES.

CHAPTER VIII.

ARTHRITIS.

A JOINT DISEASE OF THE COLT AND CALF.

Swelling of the joint of young animals soon after birth, is confined almost exclusively to the colt. I have met with but few cases in other animals, but have seen a great number of young colts die of this malady; This disease affects suckers only, consequently it might be inferred that the causes are due to an altered condition of the milk, brought about by giving non-nutritious food. I have seen colts affected, their dams having been fed on all kinds and qualities of food; hence I have come to the conclusion that we must look elsewhere for the cause, and have not arrived at any very satisfactory conclusion on this point.

SYMPTOMS.

Extreme difficulty in moving, which, perhaps, is the first and only observable sign of disease. The young animal will get up, suck and lie down again. You will soon notice a swelling of the joints, after a few days an amber colored serum —then pus begins to show itself through the skin at the joints.

The animal may live for twenty or thirty days, or longer, but recovery is very rare indeed. There is an intense fever, breathes quick, appetite impaired and thirst intense.

Curative treatment of this disease is very unsatisfactory, consequently we must turn our attention to the preventative treatment. Keep the dam in good condition, feed nothing but good nutritious food with an occasional scalded bran mash, containing from one-half to one ounce of nitrate of potash and a tablespoonful or two of common salt. During the last months that the dam carries a colt the food should be abundant and of a succulent nature; good hay, (not marsh hay and straw,) but good hay, free from mildew, moistened a few hours before feeding. Give from eight to sixteen pounds of grain a day, according to the size and make-up of the animal. It is best to work a brood mare a little every day up to a few days of foaling; at least my observations have been that those worked do better; the colts are much stronger and require less care than those allowed to remain in idleness.

DIARRHŒA, COLTS AND CALVES.

It is useless to enter into a lengthy description of this dis̄ease. There is a peculiar offensive odor to the feces, which is perhaps a thin curdle fluid. Diarrhœa is said to be contagious, as to this I am not prepared to say, but would recommend the separation of the sick from the well ones.

TREATMENT.

Give the mother two ounces of tincture of opium and two ounces of gin in one-half pint of water, as a drench. Then give the colt or calf one or two ounces of castor oil. If the diarrhœa is not checked in twenty-four hours, give the following :

Prepared Chalk............................ 1 dram,
Tincture of Opium......................... ½ "
Hydrastis Canadensis...................... 1 "

Add one egg, mix or stir well together, and give all at once, repeat this every twelve hours until relieved. It is a good practice to change the mothers food, and both mother and young should be kept in a warm, dry, comfortable place.

RETENTION OF MECONIUM—CONSTIPATION.

The meconium* is generally expelled immediately after birth, when the umbilical† circulation is first cut off. The prolonged retention of the meconium gives rise to *constipation*. This occurs more often in the colt than any other domestic animal, and is said to be the result of improper food, such as overripe hay, straw, or anything deficient in nutrative and laxative qualities.

SYMPTOMS.

A day or two after birth they become uneasy, refuse to suck, make the attempt, but nothing passes the bowels, shows symptoms of colic, rolls about, looks at its sides, back arched, and grinds its teeth together. These are the average symptoms, which may differ somewhat in the calf.

TREATMENT.

The preventative treatment consists in attending to the feeding and general health and condition of the mother before

---

* That which accumulates in the intestines before birth.
† Navel.

foaling. The young should always be fed on the first milk its parent gives. Give injections of soap and water or oil, per rectum, removing all the feces that can be reached with the finger ; repeat the injections once an hour until relieved. Give the mother a physic ball or drench (See Index), and if the injections give no relief in an hour or two give

Castor Oil.................... .... 2 ounces
Podophyllin............................... 5 grains
Fluid Extract Nuxvomica.................. 4 drops

Mix and give all at once. Twelve hours later give

Castor Oil................................ 1 ounce
Hydrastis Canadensis........ .... ½ dram

And repeat this dose once every 12 hours until relieved. If the colt will not suck, milk the mare in a warmed vessel, pour it into a warmed bottle and give it to the colt quite often.

KNUCKLING OVER.

WEAK JOINTS IN COLTS AND CALVES.

Some colts are weak and crooked legged when they first come and perhaps walk on their fetlock joint with their feet turned back.

TREATMENT.

Prepare two *Plaster of Paris Bandages* (See Index) two and one-half inches wide, then rap the cotton wadding bandages from and including the foot to the knee; have an assistant to hold the leg straight while you apply the bandages. Cover the foot with both cotton and plaster bandage as the bandage is liable to make the pastern sore if terminating above

the foot. Be sure to get the leg straight and keep it so until
the bandage drys. The bandage may be left on from one to
six weeks if there are no sores. Sores will be indicated by the
bandage getting moist or wet, in which case it must be removed at once, and the leg bathed with white lotion.

PERSISTENCE OF THE URACHUS.

LEAKING AT THE NAVEL.

During the stay of the colt in the mother the urine passes
through the *urachus* into the allantois sac. This passage is, or
should be, cut off at birth. It appears to be more frequent in
the male than the female, and is most dangerous in the male ;
weak colts are the ones most affected.

TREATMENT.

First examine the urinary organ to ascertain if the opening is all right,if not, an opening must be made. After set-
tling this point take a large pin, a small horse nail or a fine
wire nail and pass it through the skin and opening, taking a
good deep hold in the skin, then wind a wrapping twine
around the pin describing a figure 8. The pin is to be removed about the fourth day. Apply a good sized blister, using
the *fly blister* (See Index) ; do not be afraid to blister a sucking colt, as is next to an impossibility to blemish a colt with a
blister.

UMBILICAL HERNIA.

RUPTURE, OR BREACH AT THE NAVEL.

Place the colt on its back and secure it as best you can.
Then take a common wire nail two inches long (previously

ground sharp at the point) take up all the fold of the skin that hangs loose over the rupture, pass the nail through the skin and wind a good waxed end around the skin between the body and the nail; wind it tightly so as to cut off all circulation. The swelling and inflammation closes up the opening, and when the part around which the string is tied sloughs off the colt is smooth and all right. Should the first operation fail do not hesitate to operate again as there is comparatively no danger. I have often cut open the sac and sewed up the opening, but this requires some practice, and the former way is not only successful, but without danger if performed while the animal is young.

THE CALF.

Is to be operated upon in the same way as the colt, if at all, but usually it is not worth bothering with.

VETERINARY MEDICINE.

CHAPTER IX.

STANDARD REMEDIES.

These remedies are all genuine and have proven them.
selves to be just as recommended. They are no old pocket
prescriptions that " my grand father used " or that " my father
always used " but are the standard remedies of to-day formu-
lated to combat disease and assist nature in the great work she
has to perform. When nature calls for assistance, she should
have it at once, this is especially true of *all acute* diseases and
the idea of putting obnoxious drugs into the food for a sick
and debilitated animal to eat is ¡too rediculous to talk about.
Sick animals have little or no appetite and medicine placed
in their food destroys the little remaining, consequently the
craft, which is already deprived of her compass, as it were, is
now, by drugging the food, deprived of her rudder also, thus
she is left to drift for a few days, during which time she is bat-
tling with the terrific storm that is raging within. The owner
finally sends for a pilot, but alas, it is too late, she is already on
the shoal or so near that no mortal man can turn her course,
her fate is inevitable, she is a wreck, and as the storm lulls.
she sinks forever or mayhap is left stranded in her now worth
less condition, when she meets a fate worse than death by fall

ing a prey to "sharks", a class of men who call themselves "dealers." The owner of this poor wreck is anxious to realize on each and every investment so he exchanges what was once a faithful and obedient servant, (now a life wrecked by his own carelessness,) for a few paltry dollars. To such people humanity cries shame. When an animal is sick it should have medicine at stated intervals and the medicine should not be put in the food, but back over the tongue where it will be swallowed at once. By the way, I would advise those who believe in drugging horse's food, to experiment a little in this line when you are a little off and need a physic, for instance, instead of getting sugar coated pills, just take a few grains of powdered aloes and put it into your evening meal and see how much of it you will eat, and then imagine a heartless tyranical keeper who starves you until hunger forces you to eat such obnoxious stuff.

HOW TO GIVE POWDERED MEDICINE.

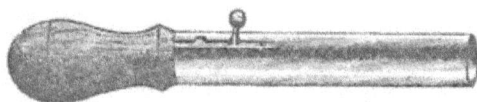

CONKEY'S DOSE GUN. [PAT. APPLIED FOR.]

Medicine should be prepared in small doses, either in the powder or liquid form, I usually prepare powdered drugs in *tablespoon doses* as you will readily see by looking over the prescriptions, and the above engraving shows an instrument which I invented for giving it. This dose gun is constructed with a spring in the cylinder so that when the thumb piece is drawn back it brings a tension on the spring, at the same time forming a cavity in the end of the gun which holds a spoonful of medicine. The instrument is carefully worked between the animals lips touching the roof of its mouth which causes it to open it, then you touch the thumb-piece and the medicine is thrown back over the tongue and swallowed. This instrument works so nicely that the task of giving medicine is mere boys-play. Sent to any address on receipt of $1. Address,

L. L. CONKEY, V. S.

6 Canal St., Grand Rapids, Mich.

GIVING LIQUID MEDICINE.

[DOSE SYRINGE.]

Large quantities of liquid medicine are to be given as a drench, which see. The stronger tinctures and extracts are best given with the one ounce syringe and should be diluted so that one syringe full of the liquid will contain a medicinal dose of the drug. The syringe is to be used the same as the dose gun.

HEAVE REMEDIES.

No. 1.

Arsenous Acid............................ 3 drams,
Powdered Willow Carbon.................. 2   "
Powdered Starch......................... 1 lb.

Mix thoroughly together and give one tablespoonful three times a day in the feed:

No. 2.

Fluid Extract of Digitalis.................... 1 ounce.
" " " Veratrum Virides.......... 1 "
" " " Aconite Root.............. 1 "
Water. .... 1 pint.

Mix and give one tablespoonful three times a day.

No. 3.

Tincture Opium........................... 4 ounces,
" Ginger........................... 4 "
" Belladonna...................... 4 "

Mix and give two tablespoonsful three or four times a day.

No. 4.

Fluid Extract Nuxvomica.................. 2 ounces,
Water..................................... 1 pint.

Mix and give one tablespoonful three times a day in the feed·

These remedies are all powerful and may take life if given in over doses, but if given as directed, no harm can possibly come from their use, on the other hand, No. 1 and No. 4 are excellent tonics.

Should you wish to shut the heaves off quickly, give one tablespoonful of No. 2, or two tablespoonsful of No. 3 once every hour until you have given three or four doses, then do not give it so often. (See heaves, page 24.)

WHITE LOTION.

Zinc Sulphate............................. 1 ounce,
Lead Acetate.............................. 1 "
Water .... 1 quart.

Mix and apply from once to five times a day, as the case may require.

YELLOW LOTION.

Take of unslacked lime, four pounds; slack it and add water to make two gallons, let it settle and pour off the water, and add one ounce of corrosive sublimate and it is ready for use. This is a splendid remedy for scratches, greese, cracked heels, etc. It is much used in the large breeding stables.

RED UTERINE LOTION.

Permanginate of Potash.................... 2 drams,
Warm Water........:..................... 2 gallons.

The water should be blood warm, and a good way to test the degree of warmth, is by introducing your hand into the uterus, carrying the rubber *injection funnel* in with it, and allow the hand to remain while the water is being poured in, should it feel cool to your hand, have it warmed, or if too warm, have it cooled.

AQUA CORROSIVE LOTION.

Corrosive Sublimate...................... 1 dram,
Boiling Water..... .... 1 pint.

Mix, and when cool, is ready for use. This is an excellent remedy for sores of all kinds, and is much used for dressing poll evil, fistulus, withers, etc. I use it for one or two dressings in all punctured and lacerated wounds.

CARBOLIZED OIL.

Cabolic Acid.............................. 1 ounce,
Sweet or Linseed Oil...... .... 1 pint.

CARBOLIZED WATER.

Carbolic Acid.............................. 1 ounce,
Water..................................... 1 pint.

CARBOLIZED VASELINE.

Carbolic Acid............................. ½ ounce,
Vaseline................................. 8 "

These remedies are used for dressing wounds and old indolent sores.

WHITE HEALING POWDER.

Prepared Chalk........................ 8 ounces.
  "   Starch......................... 8 "
Tannic Acid.............................. 1 "
Sulphate of Zinc......................... ½ "

Rub all together in a mortar and can for use. This is an excellent dressing for collar galls, greese, scratches and old indolent sores. It is equally good for man, and has often been used in sumach poisoning with favorable results.

POWDERED BORASIC ACID.

A splendid dressing for wounds and sores of all kinds. Sprinkle it over them once or twice a day.

BLACK POWDER,—STYPTIC.

Powdered Willow Carbon.................. 1 ounce,
  "   Gum Arabic..................... 1 "
  "   Rosin.......................... 1 "

Mix, and can for use —"Bartrum." This is used to stop the flow of blood. Take a bunch of cotton batting and sprinkle it over with the powder, sift a little of the powder over and into the wound, and apply the batting, keep it there by means of a bandage, or take two or three stitches through the skin, drawing the thread down over the cotton.

IODOFORM makes an "A No. 1" dressing for fresh cuts, either in man or beast, bathe with cold water until the bleeding stops, then sprinkle the iodoform over the wound.

QUININE is a fine dressing for collar galls, and may be used with advantage as a dressing in many places. Sprinkle the dry powder over the moistened wound.—"Paul."

HARTS-HORN LINIMENT.

Aqua Ammonia	2	ounces,
Spirits of Turpentine	2	"
Sweet Oil	2	"
Alcohol	2	"

Mix, and bathe the affected parts once or twice a day until you have produced a mild blister, then omit for a week and repeat. This liniment is used in all mild throat troubles, and is often used to warm up the legs in colic and congestions. It is also a good liniment for rheumatism, sprains and strains in man or beast.

TURPENTINE DRENCH.

Raw Linseed Oil	1	pint,
Spirits of Turpentine	2	ounces.

Mix and give all at once as a drench. This drench is especially adapted to *influenza* and *colic*, also to hasten the action of physic, and is intended for the horse and cow of mature years. The yearling will require one-fourth; the two-year-old, one-half; the three-year-old, three-fourths; and the four-year-old, a full dose.

LEG AND BODY WASH.

Ponds Extract of Witch-hazel............... 1 quart.
Rain Water.............................. 3   "

Mix. This is one of the best legs and body washes that has ever
been used on a speed horse, it will not blister or injure the animal in any way. It may be used much stronger without injury, and it is perhaps best to make it equal parts for old
horses.

CONDITION POWDERS.

No. 1.

Powdered Nuxvomica.................... 4 ounces,
    "   Ginger..................... 4   "
Oil Cake Meal......................... 8   "

Mix and give one tablespoonful three times a day in the feed.

No. 2.

Quinine............................. .... 2 drams.
Powdered Nitrate of Potash.............. 8 ounces.

Mix and give one tablespoonful three times a day in the feed.

COUGH BALL.

Powdered Opium........................ 1 dram,
    "   Digitalis................... 1   "
    "   Gum Camphor.............. 1   "
Calomel.............................. 1   "
Gum Arabic.......................... ½   "

Water to make into a stiff mass, to be given all at once as a
ball and repeated once a day for three days, then omit the calomel, and continue until you have the desired effect. This is
admirably adapted for the suppression of heave cough.

VETERINARY DOSE TABLE.

Giving a List of Medicines used.

Abbreviations:—Gr. Grains.   Oz. Ounces.   Dr. Drams.   M. Drops.

	HORSE.	COW.	SHEEP.	HOG.	DOG.
Aconite, Tinct.	10 to 20 M.	10 to 20 M.	2 to 3 M.		1 to 2 gr.
Alcohol	1 oz.	1 to 2 oz.	½ oz.	2 dr.	1 dr.
Aloes, Barb.	8 to 10 dr.	2 to 2 oz.	4 dr.	2 dr.	½ dr.
Alum	2 to 4 dr.	1 to 4 dr.	½ dr.	½ dr.	10 gr.
Ammonia, Carb.	2 to 4 dr.	3 to 6 dr.	½ dr.	½ dr.	6 gr.
Ammonia, Liq. Acit.	2 to 4 oz.	2 to 4 oz.			
Areca Nut	4 to 6 dr.				¼ to 1 dr.
Asafoetida	2 to 4 dr.	2 oz.			
Arsenic	5 to 8 gr.	5 to 10 gr.			
Atropine	½ to 1 gr.	½ to 1 gr.			1-30 gr.
Belladonna, Fl. Ext.	½ to 1 dr.	2 to 3 dr.	½ dr.		2 M.
Buckthorn					1 to 2 oz.
Camphor Gum	1 to 2 dr.	2 to 4 dr.	30 gr.	30 gr.	6 gr.
Castor Oil	1 pint.	1 pint.	2 to 4 oz.	2 to 4 oz.	1 oz.
Catechu	1 to 3 dr.	2 to 6 dr.			
Chloral Hydrate	½ to 2 oz.	1 to 2 oz.	1 to 3 dr.	1 to 3 dr.	15 gr.
Cod Liver Oil	2 oz.	2 to 4 oz.	1 oz.	6 dr.	2 dr.
Copper Sulphate	1 to 2 dr.	1 to 4 dr.	10 gr.	4 gr.	¼ gr.
Croton Oil		½ to 3 dr.	5 to 10 M.	8 M.	4 M.
Digitalis, Fl. Ext.	10 to 20 M.	10 to 30 M.			
Ergot of Rye, Fl. Ext.	½ to 1 oz.	½ to 1 oz.	1 dr.	1 dr.	½ dr.
Ether, Sul	1 to 2 oz.	2 to 3 oz.	2 to 3 dr.	2 to 4 dr.	30 M.
Gentian Root	4 to 8 dr.	1 to 2 oz.	1 to 3 dr.	30 gr.	15 gr.
Ginger	1 to 8 dr.	1 to 2 oz.	1 to 2 dr.	30 gr.	15 gr.
Hydrastis Canadensis	2 to 4 dr.	1 to 2 oz.	1 to 3 dr.		20 M.
Hemlock, Tinct.	2 to 3 oz.	2 to 3 oz.			2 dr.
Iodine Crystal	20 to 60 gr.	30 to 90 gr.	10 to 20 gr.		
Ipecacuan	1 to 3 dr.	1 to 3 dr.	30 gr.	25 gr.	20 gr.
Iron, Sulphate	1 to 3 dr.	2 to 4 dr.	20 gr.	10 gr.	5 gr.
Iron, Tincture Mure.	1 to 2 dr.	1 to 3 dr.			
Jaborandi	2 to 4 dr.	2 to 4 dr.			
Juniper Oil	1 to 2 dr.	1 to 2 dr.			6 M.
Lead, Acetate	1 dr.	1 dr.			
Linseed Oil, R	½ to 1 pint.	1 to 2 pints.	6 to 8 oz.	6 to 8 oz.	1 oz.
Morphia, Acetate	3 to 10 gr.	3 to 10 gr.	½ to 2 gr.		½ gr.
Nux Vomica, Powd.	1 dr.	2 to 3 dr.			
Opium, Tinct.	1 to 3 oz.	1 to 3 oz.	2 to 6 dr.	2 to 4 dr.	15 M.
Podophyllin	1 to 2 dr.	1 to 2 dr.			1 gr.
Potash, Iodide	1 to 4 dr.	2 to 6 dr.	20 gr.	20 gr.	6 gr.
Potash, Nitrate	½ to 1 oz.	1 to 2 oz.	1 to 2 dr.	30 gr.	10 gr.
Potash, Chlorate	½ to 1 oz.	½ to 1 oz.	20 gr.	20 gr.	5 gr.

	HORSE.	COW.	SHEEP.	HOG.	DOG.
Quinine	15 to 30 gr.	30 to 40 gr.	6 gr.	6 gr.	1 gr.
Root, Licorice	2 to 4 dr.	2 to 4 dr.			
Savin Oil	3 to 4 dr.	3 to 4 dr.			
Strichnine	1 to 3 gr.	2 to 5 gr.			
Salicylic, Acid	1 to 2 dr.	1 to 2 dr.			
Strophantus	½ to 1 dr.	½ to 1 dr.			
Soda, Bicarbonate	1 to 8 dr.	6 to 12 dr.	½ to 1 dr.	½ to 1 dr.	15 gr.
Soda, Hyposulphite	2 to 4 oz.	2 to 4 oz.	½ to 1 dr.	½ to 1 dr.	15 gr.
Salt, Commom	4 to 8 oz.	4 to 8 oz.	1 to 2 dr.	1 to 2 dr.	½ dr.
Salt, Epsom	4 to 16 oz.	8 to 24 oz.	1 to 2 oz.	1 to 2 oz.	2 dr.
Salt, Globber		10 to 30 oz	2 to 4 oz.	2 to 4 oz.	
Sulphur	½ to 1 oz.	1 to 2 oz.	1 to 2 dr.	1 to 2 dr.	2 dr.
S. S. Nitre	1 to 2 oz.	1 to 4 oz.	2 to 4 dr.	2 to 4 dr.	20 M.
Tannic Acid	½ to 2 dr.	1 to 3 dr.	15 gr.	15 gr.	5 gr.
Turpentine, Spts	1 to 2 oz.	1 to 2 oz.	1 to 4 dr.	1 to 4 dr.	½ dr.
Veratrum Viride. Fl. Ext.	15 to 20 M.	20 to 30 M.			
Zinc, Sulphate	1 to 2 dr.	1 to 3 dr.	15 gr.	15 gr.	2 gr.

The foregoing doses are intended for the average *horse* and *cow* of mature years. A large strong animal will require from ⅛ to ¼ more while a small one will usually require less. The weanlings ⅛, the yearlings ¼, the two-year-olds ½, the three-year-olds ¾, and the four-year-olds a full dose.

Peck Bros., Druggists, 129 & 131 Monroe St., Grand Rapids, Mich., keep a copy of this book on file, and parties wishing medicine can order it by mail or otherwise by designating page and disease. I buy of them and can recommend them to the public as selling *pure drugs* at *living prices*.

L. L. CONKEY, V. S.

CONKEY ON THE HORSE.

This book can be obtained by remitting the price, $2.00, to the author.

DR. L. L. CONKEY, V. S.,

Office with Dr. Best, M. D., No. 6 Canal Street,

Grand Rapids, Mich., U. S. A.

INDEX.

A

Asthma...................... 24
Azoturia 53
Anasarca.................... 70
Accidents.................152-108
Abscess.....................172
Age as indicated by the teeth...177
Albee, A. N.................150

B

Breathing.................... 6
Breached.................... 42
Bloat, Clover................ 48
BOTS: in the Horse.......... 64
    History of............... 65
BLOODY Murren.......... 69
    Flux................... 69
Bone Spavin................101
Bog Spavin.................125
BLISTER, The Fly..........106
    Mercurial..............106
    How to.................107
    General Remarks........107
    How often..............107
    Swelling...............108
    Accidents. To prevent.....108
    Dread of...............108
    The result of..........109
Barbed wire cuts...........122
BANDAGE: Eight tailed.......129
    Plaster of Paris..........135
Broken Legs................135
Bleeding after Castration......155

C

Congestion, Lungs........... 7
Catarrh..................50- 18
COLIC:..................35-172
    Symptoms.............. 36
    Second class........... 40
    Third   " 41
    Fourth " 43
    Fifth   " 43
    Sixth   " 45
Constipation..............40-214
Cows Cud................... 52
CATHETER: Female........ 56
    Male................... 56
Cough, Chronic............. 58
CORNS 82
    Shoeing for............ 83
Coffin Joint Disease........... 87
CONTRACTION:............. 87
    Shoeing for 88
Canker in the Foot.......... 92
Curb......................105
Cautory, Actual...........109
Cramps of the Thigh.........111
Congestion and Inflamation df'd..115
Collar Boil.................120
CAPPED Hock:.............132
    Elbow................132
Cock Ankle.................139
CHOKE: Horse.............140
    Cow..................140

CASTRATION.................142
 Age.....................143
 Abscess.................172
 Accidents...............152
 Bleeding after.......... 155
 Bull....................158
 Boar....................160
 Buck....................162
 Cryptorchide............146
 Care after..............157
 Calves..................159
 Cat.....................165
 Colic...................172
 Champignon..............173
 Clamp...................155
  " Powder..............156
  " Tongs...............157
 Dog.....................165
 Diseases following 171
 Ecraseur................154
 Flankers................147
 Gangrene................173
 Lamb....................163
 Liautard................145
 Maggots following........175
 Preparation for..........144
 Ridgling Horse..........145
  " Bull............159
  " Boar............161
 Season of...............144
 Standing................147
 Throwing for............147
COLT AND CALF DISEASES..
 Arthritis, Swelled Joints...112
 Breached................224
 Constipation............114
 Diarrhœa................113
 Leaking at the Navel......116

D

DISTEMPER. (See Influenza.). 12
 Dog.....................15
DRENCH: through the Nose.... 23
 The.....................22
Dry Murren.................. 47
Diabetes.................... 60
DIARRHŒA. Horses.......... 69
 Cattle................., ... 69
Colt and Calf...............113
Distention of the Capsular ligm't.110

Difficult Urine...............126
Dentistry. the Teeth 176
Deafness....................188

E

Eczema..................... 24
Epilepsy.................... 32
Ecraseur...................154
EYE, Inflammation of........184
 Opthalmia.................185
 Moon Blind................185
 Haw Diseased.............186
 Warts in.................187
EAR, Warts in..............188

F

Founder..................... 61
FOOT, The Horses 76
 Description of 77
 The Right................. 78
 Cracks in................. 79
 Corns...... .. 82
 Shoeing for Contration...... 88
 Seaton.................... 89
 Thrush in................. 91
 Canker.................... 92
 Nail in................... 94
 Gravel in................. 95
 Quittor in.. 96
F.RING IRON, pointed........ 99
 Feathering................106
 Bulb 138
 A Spavin.............103-104
 A Curb...................105
 Actual Cautory...........109
Fractures...................134
Fly Blown...................175

G

Greese..... 60
Gad Fly.......... ... 64
GLANDERS, Acute........... 71
 Chronic................... 73
Goiter.....................130
Gangrene...................173

H

Horse Distemper.... 16
Heaves..................... 24
Hypodermic Syringe........ 42
Hoven.......... 48

Hollow Horn.................... 50
Husk in Sheep................. 66
Hobbles, Conkey's Pat........151
HERNIA, Horse..............164
    Hog......................161

I

Inflammation, of the Lungs.... 8
INFLUENZA, Horse 12
    Dog 15
    Sheep................... 15
    Hog..................... 16
Irregular Strangles........... 17
Itch, Mane and Tail........... 28
Injection Funnel............. 43
IMPACTION Colon........... 46
    Rumen................... 47

K

Key to Practice............... 5
Kidney Disease............... 53
King H. L................... 97
Knee Bruised...............121
Knuckling...................139

L

LUNGS, Congestion........... 7
    Inflammation............. 8
Lymphangitis................. 25
Lice........................ 30
LOCK Jaw................... 33
    Body.................... 33
Laryngitis...............19–51
Losing the Cud.............. 52
Lung Worms................ 66
LAMENESS, Locating......54–75
Lithotomy..................126
Laminitis................... 61

M

Milkleg. (See Lympangitis..... 25
MANGE, Horse.............. 29
    Dog.................... 30
Megrims.................... 32
Maw bound................. 47
Merrick, Horace Big Spring,... 85
Maggots...................175
Milk Fevers................203
MEDICINE..................218
    Aqua Corrosive Lotion.......223

Borasic Acid...............223
Black Powder–Styptic.......223
Ball giving................. 20
Cough Ball................. 225
Condition Powders...........225
Carbolized Oil..............222
    "   Water............222
    "   Vaseline..........223
Drench..................... 23
    "   Turpentine...........223
    "   Through the Nose.... 22
Dose Table.................226
    "   Gun................219
    "   Syringe.............220
How to give................219
Healing Powder; White......223
Heave Remedies.............220
Iodoform dressing...........223
Liniment, Hartshorn.........223
Leg wash...................223
Quinine dressing............223
Uterine Tincture............207
    "   Red Lotion.........222
White Lotion...............221
    Yellow..................222

N

Neurotomy, Nerving......... 86
NAVICULAR Disease........ 87
    Arthritis................ 89
NEEDLE, Seaton............. 89
Nail in the Foot............: 94

O

Ossified Lateral Cartilage..... 85
Open Parotid Duct.. 124
O'Brien, P. H...............150
OBSTETRICS.   189
AFTER BIRTH retention........206
    Imprisoned 207
Abortion.. 208
Apoplexy...................203
BAG Caked.................209
    Inflammation of........210
Disowning its young.........205
EMBROYOTOMY.............195
    Instruments for..........193
    Remarks................199
PRESENTATION Breach........198

Natural..................................193
Head back..........................195
Tumultuous Labor.....................200
WOMB, Spasms of..................201
    Inflammation of.................202
    Induration, Mouth of.......201
Free Martins........................191
GESTATION, Period of........189
    Bitch.............................190
    Cat...............................190
    Cow.............................189
    Care during................192
    Mare.............................189
    Pig................................190
    Sheep and Goat............190
Twins...................................191

P

Pulse of Animals................ 5
Pleurisy............................ 9
Pluro-pneumonia.............. 11
Physic, Ball.................... 11
    Drench........................ 22
    Remarks on............... 23
    How often.................. 23
    Bad results................ 23
    Superpurgation.......... 23
Paralysis......................... 53
Purpura Hæmorhagica....... 56
Pearl Geo........................ 66
POULTICE. Oil cake Meal..... 82
    Wood Ashes............... 97
Pointing the feet.............. 87
Proud Flesh...................119
Poisoned Wounds.............122
Pockets in Wounds...........124
Poll Evil.........................137

Q

QUARTER CRACK.......... 79
    Treatment................. 80
Quittor.......................... 96

R

Respiration...................... 6
Ruminating..................... 52
RINGBONE False.............. 97
    True.............................. 98
Rheumatism.....................111
Rumenotomy....................126

Roaring...........................129
Rose, Dr. Wm...................151
RIDGLING, Horse..........145
    Bull.............................159
    Boar.............................161
RUPTURE, Horse............164
    Colt.............................165
    Hog.............................161

S

Spasms of the diaphragm...... 11
Strangles....................... 16
    "   Irregular............ 17
Surfeit.......................... 28
Sunstroke 31
Staggers......................... 32
Skin Diseases............24-28 32
Stoppage of water........... 53
Sling, The...................... 56
Scratches....................... 59
Scouring on the Road........ 68
Swelled Legs................... 70
    " Sheath............... 70
Skeleton of the Horse........ 74
SHOE, Three-quarter......... 84
    Tip............................ 88
    Spring Heeled.............. 89
Side Line and Twist...........100
SPAVIN, Bone.................101
    Particular...................104
    Bog...........................125
SWELLING 206
    After blister...............108
STIFLE Joint...................110
    Out of Joint................112
Splint, The.....................113
Sweeny 114
Surgery 115
Sewing up Wounds...........118
Speedy Cuts....................121
Scabs on Wounds..............123
Shoe Boil.......................132
String Halt.....................133
SPAYING, The Mare..........165
    Cow...........................166
    Heifer 166
    Sow...........................167
    Bitch.........................170
    Needle.......................169
Schirrous Cord................173

T

Temperature............... 5
Thermometer................ 5
Thumps..................... 11
Trismus and Tetanus........ 33
Trocar and Canula. 49
Three-quarter Shoe......... 84
Thrush........-............ 91
Twist and Side Line........ 100
Tracheotomy................ 128
THROWING HORSES........... 147
   Farmer Miles.............. 148
   Conkey 149
   Accidents in.............. 152
TEETH, The................ 176
   Conkey's Forceps.......... 183
   Milk 178
   General remarks on........ 179
   Ulcerated................. 180
   Symptoms of bad........... 180
   Wolf...................... 181
   Dressing.................. 181
   Lampas.................... 182
   Refusing to eat........... 183

Trephine................... 182

U

URINE, Profuse............. 60
   Difficult................. 126
   Bloody.................... 126

V

Vertigo.................... 32

W

Wolf in the Tail....... 47-50
WORMS, Lung, in Sheep...... 66
   Horses.................... 67
   Calves and Pigs........... 67
WOUNDS 117
   Sewing up................. 118
   Punctured 119
   Contused.................. 120
   Poisoned (Snake Bite)..... 121
   Barbed Wire............... 122
   Pockets in................ 124
   Parotid Duct.............. 124
Whistling.................. 129
Warts...................... 138

Surgical Instruments.

All Instruments recommended in this work will be sent on receipt of price, of which a list can be obtained by addressing,

DR. L. L. CONKEY, V. S.

6 Canal Street, Grand Rapids, Mich.